Energy Efficient and Reliable Embedded Nanoscale SRAM Design

This reference text covers a wide spectrum for designing robust embedded memory and peripheral circuitry. It will serve as a useful text for senior undergraduate and graduate students and professionals in areas including electronics and communications engineering, electrical engineering, mechanical engineering, and aerospace engineering.

- Discusses low-power design methodologies for static random-access memory (SRAM)
- Covers radiation-hardened SRAM design for aerospace applications
- Focuses on various reliability issues that are faced by submicron technologies
- Exhibits more stable memory topologies

Nanoscale technologies unveiled significant challenges to the design of energy-efficient and reliable SRAMs. This reference text investigates the impact of process variation, leakage, aging, soft errors and related reliability issues in embedded memory and periphery circuitry.

The text adopts a unique way to explain the SRAM bitcell, array design, and analysis of its design parameters to meet the sub-nano-regime challenges for complementary metal-oxide semiconductor devices. It comprehensively covers low-power-design methodologies for SRAM, exhibits more stable memory topologies, and radiation-hardened SRAM design for aerospace applications. Every chapter includes a glossary, highlights, a question bank, and problems. The text will serve as a useful text for senior undergraduate students, graduate students, and professionals in areas including electronics and communications engineering, electrical engineering, mechanical engineering, and aerospace engineering. Discussing comprehensive studies of variability-induced failure mechanism in sense amplifiers and power, delay, and read yield trade-offs, this reference text will serve as a useful text for senior undergraduate, graduate students, and professionals in areas including electronics and communications engineering, electrical engineering, mechanical engineering, and aerospace engineering. It covers the development of robust SRAMs, well suited for low-power multi-core processors for wireless sensors node, battery-operated portable devices, personal health care assistants, and smart Internet of Things applications.

Energy Efficient and Reliable Embedded Nanoscale SRAM Design

Bhupendra Singh Reniwal,
Pooran Singh,
Ambika Prasad Shah and
Santosh Kumar Vishvakarma

CRC Press is an imprint of the
Taylor & Francis Group, an **informa** business

First edition published 2023
by CRC Press
2385 NW Executive Center Drive, Suite 320, Boca Raton, FL 33431

and by CRC Press
4 Park Square, Milton Park, Abingdon, Oxon, OX14 4RN

CRC Press is an imprint of Taylor & Francis Group, LLC

© 2023 Bhupendra Singh Reniwal, Pooran Singh, Ambika Prasad Shah and Santosh Kumar Vishvakarma

Reasonable efforts have been made to publish reliable data and information, but the author and publisher cannot assume responsibility for the validity of all materials or the consequences of their use. The authors and publishers have attempted to trace the copyright holders of all material reproduced in this publication and apologize to copyright holders if permission to publish in this form has not been obtained. If any copyright material has not been acknowledged please write and let us know so we may rectify in any future reprint.

Except as permitted under U.S. Copyright Law, no part of this book may be reprinted, reproduced, transmitted, or utilized in any form by any electronic, mechanical, or other means, now known or hereafter invented, including photocopying, microfilming, and recording, or in any information storage or retrieval system, without written permission from the publishers.

For permission to photocopy or use material electronically from this work, access www. copyright.com or contact the Copyright Clearance Center, Inc. (CCC), 222 Rosewood Drive, Danvers, MA 01923, 978-750-8400. For works that are not available on CCC please contact mpkbookspermissions@tandf.co.uk

Trademark notice: Product or corporate names may be trademarks or registered trademarks and are used only for identification and explanation without intent to infringe.

ISBN: 978-1-032-08159-5 (hbk)
ISBN: 978-1-032-10059-3 (pbk)
ISBN 978-1-003-21345-1 (ebk)

DOI: 10.1201/9781003213451

Typeset in Times LT Std
by Apex CoVantage, LLC

Contents

Acknowledgment ix
Preface.. xi
Author Bios xiii

Chapter 1   Introduction 1

        1.1   Memory Hierarchy.. 1
        1.2   Why SRAM in Digital Systems.. 2
        1.3   Low-Power SRAM Architecture 4
                1.3.1   SRAM Design Trade-Offs 6
        1.4   Data Read Operation in SRAM.. 10
        1.5   Mismatch Issues in SRAM and Failure 10
                1.5.1   Read Failure 11
                1.5.2   Write Failure 12
                1.5.3   Data Retention Voltage 12
                1.5.4   Bitline Capacitance Mismatch 13
        1.6   Reliability Failure Mechanisms....................................... 15
                1.6.1   Time-Zero Variability 15
                1.6.2   Time-Dependent Variability 18
                1.6.3   Bias Temperature Instability 20
        1.7   Soft Errors..24
        1.8   Fin-Shaped Field-Effect Transistors and Alternative
               Technologies for Reliable SRAM Design.................................26
                1.8.1   Fin-Shaped Field-Effect Transistors 26
                1.8.2   Silicon on Insulator 27
                1.8.3   Tunnel Field-Effect Transistors.........................29
        1.9   Applications of ULP SRAM...30
                1.9.1   Application of SRAM in Look-Up Table of FPGA 31
                1.9.2   Application of SRAM in IoT Edge Devices...............33
                1.9.3   Application of SRAM in Image-Processing
                         Memory System 35
        1.10 Design Challenges in SRAM...36
                1.10.1 Design Techniques 36

Chapter 2   Design Metrics for Embedded SRAM 45

        2.1   Conventional SRAM Architecture: 6T and RD8T
               SRAM Bitcells... 46
                2.1.1   6T SRAM Read Operation 47
                2.1.2   Write Operation in SRAM 52
                2.1.3   Latch Mechanism in SRAM 55

	2.2	Bitcell Stability Metrics	56
		2.2.1 Method of Evaluating NM	58
		2.2.2 Read Static Noise Margin	60
		2.2.3 Write Static Noise Margin	63
		2.2.4 Wordline Write Trip Voltage	64
		2.2.5 Bitline Write Trip Voltage	65
		2.2.6 Dynamic Read Stability	65
		2.2.7 Dynamic Write Stability	67
	2.3	Decoders	67
	2.4	Read Driver	68
	2.5	Write Driver	70
	2.6	Leakage Power in SRAM	71
	2.7	Dynamic Power in SRAM	73
	2.8	Read/Write Latency	73
Chapter 3		SRAM Bitcells over Conventional Memories	76
	3.1	Introduction	76
	3.2	SRAM in Subthreshold FPGA	77
	3.3	PFC10T for Look-Up Table	78
		3.3.1 Architecture of PFC10T SRAM	79
		3.3.2 SRAM with Feedback Control Circuit	80
		3.3.3 Read Operation in PFC10T SRAM	83
		3.3.4 Write Operation in PFC10T SRAM	84
		3.3.5 SRAM in Standby Mode	84
		3.3.6 Design Metrices of PFC10T SRAM	84
		3.3.7 Discussion on Post-Layout Simulation Results	90
		3.3.8 Application of PFC10T SRAM in FPGAs	90
	3.4	Subthreshold Process Tolerant 10T SRAM for IoT Applications	94
		3.4.1 PT10T SRAM Bitcell	94
		3.4.2 Operations of PT10T SRAM	95
		3.4.3 PT10T Write and Read Analysis	99
		3.4.4 PT10T Leakage Power	100
		3.4.5 PT10T RSNM	101
		3.4.6 PT10T WSNM	101
		3.4.7 8-kb Subthreshold SRAM for IoT Applications	102
	3.5	Robust Subthreshold 8T SRAM for Image Processing	103
		3.5.1 PFC Subthreshold 8T SRAM	105
		3.5.2 SRAM Layout and Macroblock	106
		3.5.3 Operations of PFC8T SRAM	107
		3.5.4 Analysis of PFC8T SRAM	109
		3.5.5 Results Summary of 2-kb Array	113
		3.5.6 Application of PFC8T SRAM on an Object Tracking System	114

Contents   vii

                      3.5.7   Object Tracking Analysis and SRAM Utilization: Results and Comparisons 120

Chapter 4   Offset Correction in the Sense Amplifier 125

            4.1   Introduction 125
            4.2   Data Sensing Methods in Embedded Memories 126
                  4.2.1   Voltage Mode Approach 127
                  4.2.2   Current Mode Approach 129
            4.3   Design Metrics for Efficient Sensing Amplifier Design 130
                  4.3.1   Sensing Delay and Read Latency 130
                  4.3.2   Input-Referred Offset in Sensing Amplifiers 131
                  4.3.3   Read Yield Issue 132
            4.4   Variability Consequences on Sense Amplifiers 133
                  4.4.1   Differential Resistance Modeling of CLSA 134
                  4.4.2   Self-Correcting Sense Amplifier 137
            4.5   Discussion of Important Figures of Merits 142
                  4.5.1   Offset Measurement 143
                  4.5.2   Performance Analysis 145
            4.6   Area Consideration 150
            4.7   Chapter Summary 151

Chapter 5   Data Sensing in SRAM: A Hybrid Approach with FinFET 153

            5.1   Introduction 153
            5.2   Design Bottlenecks in Current Mode Sensing in a Scaled Complementary Metal-Oxide Semiconductor Process 154
            5.3   Differential Current Mode SA Design 154
            5.4   Differential Current Feed Techniques 156
            5.5   Capacitance Modeling of Sensing Circuit 157
            5.6   Functionality and Differential Voltage and Current Analysis 161
            5.7   Device and Capacitance Mismatch in CF-SA 163
            5.8   Offset Analysis 168
            5.9   Performance and Power Measurement 170
            5.10   Chapter Summary 172

Chapter 6   BTI-Aware and Soft-Error-Tolerant SRAM 176

            6.1   Introduction 176
            6.2   Radiation Hardening Analysis Methodology 177
                  6.2.1   Critical Charge 177
                  6.2.2   Soft-Error Rate 179
            6.3   Stable and Reliable SRAM Cells 181
                  6.3.1   Radiation-Hardened Asymmetric 10T SRAM Cell 182

	6.4	Leakage Current Estimation	183
	6.5	SRAM Bitcell Design Metrics	185
		6.5.1 Stability Analysis	185
		6.5.2 Power Dissipation Analysis	185
		6.5.3 Delay Analysis	187
		6.5.4 Critical Charge Analysis	188
		6.5.5 Aging Effects on SRAM Cells' Performance	191
		6.5.6 Soft Error Analysis	192
		6.5.7 Effect of Supply Voltage and Temperature Variation Analysis	194
		6.5.8 Process Variation Analysis	195
		6.5.9 Area Comparison	199
		6.5.10 Reliability and Stability–to–Energy Area Product Ratio	200
	6.6	Summary	201
	6.7	Glossary	201
	6.8	Questions	202
Index			205

Acknowledgment

First and foremost, we are grateful to Almighty God for blessing us with this opportunity to write this book. It gives us an immense debt of gratitude to express our sincere thanks to IBM Memory Design Team, Intel Mixed Signal Solution Group, and Programmable Solutions Group for the excellent technical discussion, constant encouragement, and patience during the writing phase. We would like to thank many colleagues at the Nanoscale Devices VLSI Circuit & System Design Group at IIT Indore; the Frontier Electronics Laboratory at IIT Jodhpur; the Integrated Circuit Reliability, Security, and Quality Laboratory at IIT Jammu; and Advanced VLSI Design Lab at Mahindra University, Hyderabad, India, who have been working on the various projects with us. We owe a lot to our families, who encouraged and helped us at every stage of our personal and academic lives and have longed to see this achievement come true. We would like to thank the High-Performance Low-Power laboratory at the University of Virginia USA for technical insight on memory design aptitude.

Preface

Continuing progress and integration levels in silicon technologies make possible complete end-user systems consisting of multicores on a single chip targeting either embedded or high-performance computing. To relentlessly meet this ubiquitous need recently evolved multicore architectures need large, embedded memories to support high bandwidth and computational requirements. Therefore, for the foreseeable future, static random-access memory (SRAM) will likely remain the embedded memory technology of choice for many microprocessors and systems on chips (SoCs) due to its speed and compatibility with standard logic processes. For these reasons, it is considered to be highly promising and attracted considerable research attention toward optimizing its reliability, power, and speed.

Consequently, designing energy-efficiency SRAMs is one of the key components for energy-efficient systems. However, without new paradigms of energy-efficient designs, producing embedded SRAM capable of meeting the computing storage and communication demands of the emerging applications will be unlikely.

Access time and power consumption of memories are largely determined by sense amplifier (SA) design. If the SA-enable (SEN) signal is asserted early, the SA cannot amplify the small voltage difference correctly. The overhead of access time and power consumption is increased if the SEN is asserted late. Therefore, the optimum timing for the SEN signal is critical for a high-speed and low-power SRAM design. The minimum required bitline swing is often limited by the SA offset voltage (V_{OS}). The V_{OS} is the minimum voltage difference between the bitlines, effective to cause the output signal to rail to a specified state. Therefore, if higher than the offset of the SA, then the required differential voltage will be higher, resulting in a higher bitline swing (more power) and delayed SEN enabling (slow memory).

With technology scaling toward the physical limit, process variations are becoming a growing concern in SA designs. Because continued process scaling tends to cause increases in the input-referred offset of SA topologies. This is largely due to the overall increase in local (e.g., within-die) process variation due to mechanisms such as lithographic variation and random doping fluctuation (RDF). These local variations cause the threshold voltage (V_{th}) of transistors with identical layout to be distributed normally, and the standard deviation of the V_{th} distribution is proportional to $1/(WL)^{1/2}$.

Therefore, nanoscale technologies unveiled two significant challenges to the design of high-speed and reliable SRAMs. The first challenge is process variation, which threatens the reliability by affecting the sensing circuit's sensitivity. This effect demands larger signal magnitudes, which decrease the speed as well as power consumption. The second challenge is the variation of the cell current (I_{CELL}), which reduces the worst-case cell current that drives the bitline. In effect, this reduction demands a longer wordline activation time to ensure sufficient bitline voltage swing for correct sensing. The wordline activation time directly influences the cell data stability. A long wordline activation time can affect the noise margin. Indeed, as V_{DD} is scaled down, decreasing memory cell current issues concerning differential current degradation in the current SA become severe.

Author Bios

Bhupendra Singh Reniwal received a PhD from IIT, Indore, India, 2016. He is currently working as an assistant professor in the Department of Electrical Engineering, Indian Institute of Technology Jodhpur, India. Post PhD, he has worked as a senior product development engineer; a semiconductor vertical in UST Global Bangalore; India Mixed-Signal IP Solution Group (MIG) at Intel Corporation Penang, Malaysia; and Systems & Technology Group, ASIC Foundry, IBM Bangalore, where he was involved on developing energy-efficient memory architecture, I/O circuit design and its pre-silicon validation for Internet of Things (IoT) applications in sub-nanometric trigate fin-shaped field-effect transistor (FinFET) processes. At IBM, he was involved in research and development on low-power methodology definition at the Schematic2GDS level for sub-nanometric nodes, especially for FinFET memory design. He has served the Department of Electronics & Communication Engineering IIITDM Kancheepuram and BITS Pilani and as an assistant professor from November 2019 to October 2022 and from May to December 2017, respectively. He is a recipient of the prestigious SIRE-2022 Faculty Fellowship from the Department of Science & Technology (DST) GOI and joined the University of Virginia, USA, as a visiting faculty. He received the User Design Best Research Paper Award at IEEE 29th International Conference on VLSI Design and Best Poster Presentation Award for Ultra Low Power SRAM Design at the Ramanujan Conclave 2016. He is a recipient of the International Travel Award as early recognition in solid-state circuit design from the Association of Computing Machinery (ACM), New York, USA, and Department of Science & Technology (DST).

Dr. Pooran Singh is an assistant professor in the Department of Electrical and Computer Engineering at Mahindra University (MU) École Centrale School of Engineering. Dr. Pooran graduated with a PhD from the Department of Electrical Engineering, IIT Indore. He is a Fulbright-Nehru Doctoral Fellow (2014–2015). Under a Fulbright Fellowship, he was associated with the Department of Electrical and Computer Engineering, Georgia Institute of Technology, Atlanta, USA, for 1 year. Prior to joining MU, he was an analog design engineer (SRAM Design) at Intel Microelectronics, Penang, Malaysia. During his time there, his primary work included designing SRAM circuits, pre-layout SRAM design, and analyzing its various design parameters, that is, read margin/write margin, critical path, read/write performance, dynamic, and leakage power at different PVT values for Intel's 7-nm and 10-nm field programmable gate arrays. His primary research work includes designing low-power and robust SRAM for space applications, in-memory computing, and Internet of Things devices.

Dr. Ambika Prasad Shah is currently working as an assistant professor in the Electrical Engineering Department and an associate dean of Corporate Relations at the Indian Institute of Technology (IIT) Jammu, India. He received a PhD degree from the Electrical Engineering Department, Indian Institute of Technology Indore,

India. Before joining IIT Jammu, Dr. Shah worked as a postdoctoral fellow at the Institute for Microelectronics, TU Vienna, Austria. He is the recipient of the Young Scientist Award from the M.P. Council of Science and Technology Bhopal, M.P. India. He has authored/co-authored more than 70 research papers in peer-reviewed international journals and conferences. He was the conference organizing chair for VLSI Design and Test (VDAT) 2022 and a fellowship chair for International Conference on VLSI Design 2022. He is a fellow of IETE, and senior member of Institute of Electrical and Electronics Engineers (IEEE).

His current research interest includes reliability analyses of digital circuits, design for reliability, fault-tolerant circuits, reliability modeling, low-power high-performance circuit designs, and hardware security circuits.

Dr. Santosh Kumar Vishvakarma is a professor in the Department of Electrical Engineering, Indian Institute of Technology Indore, MP, India. He is engaged with teaching and research in the area of energy-efficient and reliable static random-access memory (SRAM) design, enhancing the performance and configurable architecture for deep neural network (DNN) accelerators, SRAM-based in-memory computing architecture for edge artificial intelligence, reliable and secure designs for Internet of Things applications, and designs for reliability. Prof. Vishvakarma is the reviewer of various journals like *IEEE Transactions on Electron Devices*; *IEEE Transactions on Nanotechnology*; *IEEE Transactions on VLSI Integration System*; the *Microelectronics Journal* (Elsevier); the *Integration the VLSI Journal* (Elsevier); *Analog Integrated Circuits and Signal Processing* (Springer); *Circuits, Systems & Signal Processing*; and *Solid State Electronics*, among others. He is a member of IEEE, a professional member of the VLSI Society of India, an associate member of the Institute of Nanotechnology, and a life member of the Indian Microelectronics Society, India.

He was the general chair the 23rd International Symposium on VLSI Design and Test (VDAT 2019), July 4–6, 2019, IIT Indore, India.

Prof. Vishvakarma did schooling in Gorakhpur itself and then received a Bachelor of Science in Electronics, a Master of Science in Electronics, and a Master of Technology in Microelectronics from the University of Gorakhpur, Devi Ahilya Vishvidayalaya Indore, and Panjab University Chandigarh in 1999, 2001, and 2003, respectively. Dr. Vishvakarma obtained a PhD degree on the topic "Analytical Modeling of Low Leakage MGDG MOSFET and its Application to SRAM" from Microelectronics and VLSI Group, Department of Electronics and Computer Engineering, Indian Institute of Technology, Roorkee, in 2010 and worked under the supervision of Professor S. Dasgupta and Professor A. K. Saxena in the area of metal-oxide semiconductor device modeling and SRAM circuit design.

1 Introduction

1.1 MEMORY HIERARCHY

Moore's law of scaling has been the most important driving force behind the semiconductor industry for many decades of technological advancements [64]. The drive to meet the requirements of "Moore's law" also had synergistic benefits due to the physics of semiconductor operation, which caused the dimensions of the transistor gate thickness and channel length to shrink to reduce the voltage and power required for reliable operation with more functionality and performance [101]. This "triumvirate" of functionality, higher performance, and lower power market benefits to consumers, continuously required low energy on chip-embedded memory. Therefore, emerging memory technologies continue to make steady progress towards product intercepts, including phase change random-access memory (PCRAM) and resistive RAM (ReRAM), while spin transfer torque magnetic RAM (STT-MRAM) is becoming a strong candidate for both stand-alone and embedded applications [43]. However, as per the semiconductor industry perception embedded static random-access memory (SRAM) continues to be a critical technology enabler for a wide range of applications from high-performance computing and graphics to mobile applications, wearable electronics, and memory-centric systems for artificial intelligence applications [1]. This is due to SRAM having the low read latency with a low-power operation as identified in Table 1.1, which shows the performance metrics of state-of-the-art memory technologies. This is due to the fact that SRAM is one of the fastest memories using silicon that has ever produced with a low-power operation, as can be seen in Table 1.1.

Accordingly, to benefit efficiently from transistor scaling, modern digital architectures increasingly emphasize the use and integration of more and more SRAMs [95,75]. The resulting consequence in modern systems on chip (SoCs) is that embedded SRAM occupies a significant portion of SoCs and has a large impact on chip yield. With this motivation to increase the capacity of on-chip cache, SRAM is predicted to occupy about 90% of die area by 2020 [92].

In addition, Table 1.1 shows the recent development in memory technology, and Figure 1.1 shows the percent of the die area used for embedded SRAM in advanced-performance multicore SoC [2]. It can also be determined from Figure 1.1 that the allocation of physical real estate (die area) of typical large application specific integrated circuit (ASIC) and SoC designs tends to fall into three general groups. Die area dedicated to new custom logic (new area logic). Die area dedicated to reusable logic (3rd-party IP or legacy internal IPs). Die area used for embedded memory. It affirms that the intensive data processing for energy-efficient computation requires larger on-chip energy-efficient SRAMs [2].

TABLE 1.1
Comparison of Performance Metrics of Various State-of the-Art Memory Technologies

Memory Features	Volatile		Non-Volatile			
	DRAM	SRAM	FLASH	MRAM	STTRAM	PCRAM
Area (F2)	8-12	50-80	4-11	6-20	8-25	4-6
Write Time	6-80 ns	2-50 ns	10-200 μs	3-30 ns	3-30 ns	20-100 ns
Read Time	6-80 ns	2-50 ns	60ns-20μs	3-30 ns	2-20 ns	20-50 ns
Bits/cell	1	1	2-4	1	1	1-2
Read Endurance	$>10^{16}$	$>10^{16}$	10^{16}	$>10^{15}$	$>10^{15}$	10^{8}-10^{14}
Write Endurance	$>10^{16}$	$>10^{16}$	10^{6}	$>10^{15}$	$>10^{15}$	$>10^{12}$
Destructive Read	Yes	Yes	No	No	No	No
Scalability	Capacitor	6-Transistors	Tunnel Oxide	Current Density	Current Density	Litho

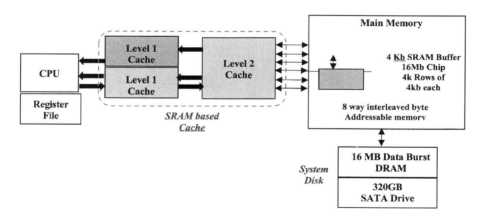

FIGURE 1.1 SRAM-based cache memory in a digital system.

1.2 WHY SRAM IN DIGITAL SYSTEMS

Performance, power management, reliability, and cost-effectiveness are the four major entities of a digital integrated circuit (IC) in very large-scale integration (VLSI) systems. Almost all digital electronic devices are battery-enabled and manufactured using semiconductor devices. The battery-enabled device requires a long

Introduction

power backup to perform multiple tasks associated with smart technology equipment. Over the past few decades, the technology has been scaled down from deep submicron to nanometer technology, which leads to higher leakage current and hence increases power consumption. The density of the number of transistors per unit area is also increasing, which reflects higher power dissipation [44]. In addition, the performance in terms of the speed of a digital IC is directly proportional to its power dissipation [13]. It means that for a high-performance device, users have to deal with its higher power consumption. In addition, due to the constant scaling of complementary metal-oxide semiconductor (CMOS) devices, the issues related to the yield and reliability of ICs are also growing. The yield of a digital CMOS circuit depends on its power-delay product. In high-speed devices, equivalent power is increasing with the technology scaling and eventually affects the yield. While concentrating on low-power devices there is a limitation on speed. Therefore, to obtain a better yield, the focus should be on a certain window that offers a balance between speed and power in nanometer technology [36].

In digital systems, SRAM is considered as the key component to store and read binary information. A conventional SRAM cell is designed by a pair of two cross-coupled inverters and two n-type metal-oxide semiconductor (NMOS) access transistors. These transistors provide entry to the cell in the read and write operations and keep storage nodes intact by separating bitlines (*BL*s) from the storage nodes during the no operation. An SRAM cell is intended to deliver nondestructive read access and write competence and data storage (or data retention) until the cell is powered. Technology trends have resulted in static and dynamic power dissipation and emerged as a primary design reflection in microprocessors. However, to keep dynamic power dissipation to a lower level, consecutive technology generations are depending on dropping the supply voltage. Additionally, in order to keep better performance, consistent reduction in the transistor threshold voltage is required [58,65]. Since the leakage current increases exponentially with the drop-in threshold voltage, the static power dissipation increases to a substantial portion of the overall chip power dissipation in contemporary nanometer processes.

Nowadays, microprocessor-controlled handheld devices are composed of embedded memory, which characterizes a huge share of the SoC. These handheld systems require ultra-low-power (ULP) circuits to operate with the battery for a longer duration. Applications of ULP SRAM are extremely broad, including a neural signal processor, subthreshold processor, biomedical implants, wireless sensing, low-voltage cache operation, and the like [58,65,105,30]. These applications demand careful design by maintaining the associated trade-off between power and speed. In order to adhere to intense scaling trends, SRAM design is also highly constrained. Moreover, parameter variations in metal oxide semiconductor field effect transistors (MOSFETs) and system-level power consumption increase the design challenges. Since the impact of SRAM on the whole processing unit is very significant, modern ULP SRAMs must be developed with their own trade-offs. The trade-offs for SRAMs are mainly related to power, performance, and density constraints as shown in Table 1.2. The SRAM design involves trade-offs related to speed and power in order to accomplish specific requirements. The power consumption in SRAM can be reduced by using new

TABLE 1.2
Applications of SRAM and Design Trade-Offs

	Ultra-low power SRAM	High-performance SRAM	High-density SRAM
Application	Biomedical implants, wireless sensing, Internet of Things edge devices	High-end servers, complex computing Image processing	Mobile, multimedia gadgets
Design Approach	High-V_{th} MOSFET low-power supply, medium-sized bit-cells, short bitlines	Low-V_{th} MOSFET large-sized bitcells, short bitlines	High-V_{th} MOSFET, small-sized bitcells, long bitlines

circuit topologies and improving the memory architecture. Moreover, stable and ULP on-chip memory is essential to attain higher reliability and longer battery life for portable applications.

SRAM-based cache memory is a very fast memory that acts as a buffer or intercessor between the central processing unit (CPU) and main memory as shown in Figure 1.1. It temporarily stores the frequently needed data and instructions so that they are instantly available to the CPU whenever required. However, SRAM-based cache offers a very high speed for data transfer; it is relatively expensive compared to other memory technologies in an electronic system. A major purpose of the cache is to reduce the memory access time. Cache has two main sections. One is the directory, which stores the address, and second is data lines, which particularly stores the data, which are later addressed by the directory. The cache controller controls the data transfer between the cache memory and the CPU. In the event of, the requested data element is not present in the cache, the cache controller calls for that data from memory. Thus, the read and write requests to memory are handled by the memory controller. If the data requested by the CPU are available in the cache, it is fetched and sent to the processor by the cache controller, and this is considered a cache hit. If the data requested by the CPU is not present are the cache, this is considered a cache miss. The throughput of a system mainly depends on the speed of cache hits because in this case, the CPU has to face a very short latency [54]. Consequently, SRAM becomes the mainstream technology for first- and second-order memory as a cache. This enables the systems to execute the operations with low latency for high-performance applications.

1.3 LOW-POWER SRAM ARCHITECTURE

A typical SRAM memory architecture along with a simple schematic illustration of column circuitry is shown in Figure 1.2. The data storage element, or core, consists of 6-transistor (6T) memory cells (shown in Figure 1.3a) organized in an array of rows and columns. Here, each bitcell is capable of storing one bit of binary data. Also, each 6T bitcell shares a shared connection with other cells in

Introduction

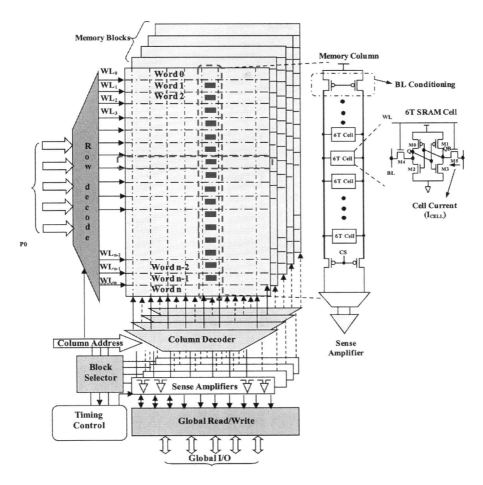

FIGURE 1.2 SRAM memory architecture view.

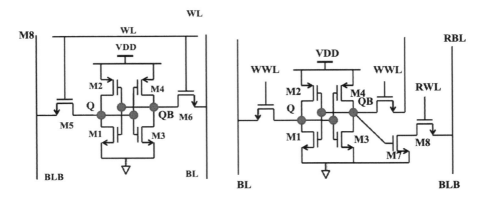

FIGURE 1.3 SRAM cells: (a) A conventional 6T SRAM and (b) a standard read-decoupled 8T SRAM.

the same row called wordline (WL) and an additional common signal with the other cells in the same column called column select (CS). In this structure, there are 2^i rows and 2^j columns. Thus, the total number of memory cells in this array is $2^i \times 2^j$.

The 6T SRAM cell shown in Figure 1.3a contains a pair of cross-coupled inverters (M1–M4), which holds the data, and a pair of access transistors (M5–M6) to initiate the read or write operation. However, a read decoupled 8T shown in Figure 1.3(b) is designed to isolate the read–write path and to improve the static noise margin (SNM) in the read operation. To access a particular memory cell, the corresponding CS and WL must be activated. The row and column decoders are employed to accomplish row and CS operations, respectively. The data are written into the SRAM bitcell by applying the logic bit value and its complemented value onto the bit lines and then triggering the wordline (WL for 6T and write wordline (WWL) for 8T). In addition, the column circuitry also consists of the global read/write circuitry, the *BL* sensing circuitry, and the column multiplexers as shown in Figure 1.2.

1.3.1 SRAM Design Trade-Offs

Designing SRAM for low-power applications and high-performance computing has its own challenges and trade-offs. An application-specific SRAM design is designed for various modes of operations like low-power devices, high-performance devices, and high-reliability devices. The three major trade-offs for designing SRAMs in subthreshold regions are power, speed, and noise margins.

1.3.1.1 Power Consumption in SRAM

With the rapid development of portable digital applications, the demand for a larger block with high-performance embedded SRAM has increased. Consequently, SRAM exhibits a dominating portion of the total die area, which results in the largest share of power consumption in a system. Consequently, SRAM power dissipation is widely recognized as a top-priority issue for the research community. The challenges widely faced by the research community are to find an alternate structure of a transistor, design new architecture, and effectively apply circuit techniques that can balance the need for high-performance SRAM devices with a low-power dissipation.

One of the most effective methods to decrease the power consumption and thus extend the battery life is lowering the supply voltage. However, operations at reduced voltages degrade robustness due to low noise margins and higher vulnerability to process variations and device mismatches. This is why the conventional SRAMs are usually designed to function vigorously with a minimal supply voltage. Due to the ratio design of the conventional 6T SRAM, the operational margins move away from the nominal point, the effective margins that are essential for the functionality start to reduce. Furthermore, local and global transistor variations tied with aging effects limit the design space for SRAM functionality. Subsequently, it becomes tough to make an SRAM operative at lower supply voltages. In addition, the key factors of power consumption in sub-nanometer SRAM cells are discussed in the following subsections.

Introduction

1.3.1.1.1 Dynamic Power Dissipation

There are three ways by which dynamic power dissipation can occur, namely, switching power, short-circuit power, and glitching power.

Switching power: It occurs due to the charging and discharging of parasitic capacitances across the *BL*s of SRAM. The average energy from V_{DD} to the *BL* capacitance (C_{BL}) to charge the capacitor across the storage node is equal to $C_{BL} \times V_{DD}^2$. Hence, the total switching power dissipated in one cycle of input pulse can be computed as

$$P_{\text{switching}} = \frac{1}{T} \times C_{BL} \times V_{DD}^2 = f_{clk} \times C_{BL} \times (V_{DD})^2. \tag{1.1}$$

Short-circuit power: Short-circuit current happens during transitions of signal when both the NMOS and p-type metal-oxide semiconductor (PMOS) of SRAM inverters are ON and there is a short circuit path between V_{DD} and the ground of the cross-coupled inverters. The power dissipates due to this current is known as short-circuit power dissipation. The short-circuit power accounts for more than 20% of total power dissipation. Short-circuit power dissipation increases with clock frequency, due to the increase in the number of transitions. The short-circuit power is determined by

$$P_{SC} = I_{mean} \times V_{DD}, \tag{1.2}$$

where I_{mean} is the average current passes through V_{DD} to ground (*GND*) in a short-circuit condition.

Glitching power: Glitches in the circuit are the undesired signal switching, which do not hold any valuable data. However, it contributes to switching and short-circuit power dissipation. Glitches can be divided into two classes namely, generated and propagated. If the input to a gate is skewed in time, then there is a chance of generating a glitch at the output. If a glitch arrives at the input of the gate when it is active, a propagated glitch will occur. The number of glitches in an SRAM depends on the logic function and the number of inputs to the gates of the transistors associated with SRAM. Generally, a 6T SRAM has only two inputs, the WL and the *BL*. Hence, the glitching power occurs only due to the timing mismatch between WL and *BL* inputs. In the case of a standard 8T SRAM, there is one more input, that is, read wordline (RWL), which can cause a further increase in glitching power.

1.3.1.1.2 Static Power Dissipation

It becomes active when the SRAM is in a hold state, that is, when SRAM is holding data in storage nodes. The static power dissipation comprises subthreshold and reverse-biased diode leakage currents. Due to the scaling of threshold voltages, the subthreshold leakage power becomes more dominant.

1.3.1.1.3 Sensing Power

It is associated with the read operation of the SRAM. An SRAM requires a differential SA to sense a small *BL* differential voltage (ΔV_{BL}) and read the bit information stored in the SRAM. Therefore, the higher the value of sensing offset voltage (ΔV_{BL}), the more sensing power is required.

1.3.1.2 Limitations of Subthreshold SRAMs

To reduce both static and dynamic power dissipation in SRAM, the most effective technique is to scale down the supply voltage. However, it has a few limitations, such as low SNM and current variations due to process discrepancies [58,65,105,30]. The subthreshold operation can be attained by setting the supply voltage below the threshold voltage (V_{th}) of the MOSFET. Scaling of the supply voltage below the device threshold shows the exponential dependence of the drain current on the gate voltage. The power consumption can be reduced by operating the MOSFET in the subthreshold regime. The basic equation of subthreshold current and total off current [36] of the MOSFET is as follows:

$$I_{ds} = I_{ds0} \times e^{\frac{V_{gs} - V_{th0} + \eta V_{ds} - k\gamma V_{sb}}{\eta V_{th}}} \left(1 - e^{\frac{-V_{ds}}{V_{th}}} \right) \quad (1.3)$$

$$I_{ds0} = \mu_0 C_{ox} \frac{W}{L} \times (\eta - 1) \times V_{th}^2, \quad (1.4)$$

where V_{th0} is the transistor threshold voltage for zero substrate bias, η is the subthreshold slope factor $\left(\eta = 1 + \frac{C_d}{C_{ox}} \right)$, C_d is the drain capacitance, C_{ox} is the oxide capacitance, I_0 is the off current at $V_{gs} = V_{th}$, V_{ds} is the drain to the source voltage, V_{gs} is the gate to the source voltage, V_{sb} is the source to the body voltage, and μ_0 is the mobility.

The threshold voltage is

$$V_{th} = V_{th0} + \gamma \left(\sqrt{\phi_s + V_{sb}} - \sqrt{\phi_s} \right), \quad (1.5)$$

where the surface potential at the threshold is

$$\phi_s = 2V_{th} \ln \frac{N_A}{n_i}. \quad (1.6)$$

The body effect coefficient is

$$\gamma = \frac{t_{ox}}{\varepsilon_{ox}} \sqrt{2q\varepsilon_{si}N_A} = \frac{\sqrt{2q\varepsilon_{si}N_A}}{C_{ox}}. \quad (1.7)$$

It is noted from the preceding equations that I_{ds} is measured as the basic circuit design parameter. As $I_{ds} \propto \dfrac{W}{L}$, the transistor sizing aspect ratio W/L is not so effective in changing I_{ds}. However, the threshold voltage variation can be very effective in changing I_{ds} when designing a subthreshold SRAM circuit. Gate current due to carrier tunneling through the oxide is negligible compared to I_{ds}. Also, the junction leakage current is negligible in the subthreshold regime related to I_{ds}.

1.3.1.3 SRAM Stability

The SRAM stability is evaluated using two key factors namely, static noise margin (SNM) and write trip point (WTP). The SNM also has three key factors associated with an SRAM cell, that is, read SNM (RSNM), hold SNM (HSNM), and write SNM (WSNM). The RSNM/HSNM can be evaluated by plotting the largest possible square inside the two voltage transfer curves (VTC) of the CMOS inverters, as shown in Figure 1.4. The RSNM is defined as the length of the side of a square, given in volts. When an external DC noise is larger than the RSNM/HSNM, the state of the storage node of SRAM will change and the stored data will be lost or corrupted. The WSNM is measured by plotting the VTC of the write '1' operation by sweeping QB from 0 to V_{DD} and VTC of the write '0' operation by sweeping QB from V_{DD} to 0. The WSNM is measured as the smallest square plotted into the VTC curve as shown in Figure 1.4. When the external noise is greater than that of the WSNM, the SRAM faces a write failure.

The other key factor of stability in SRAM is WTP. The WTP is observed at the write operation when an opposite noise is applied to the storage node. The WTP is measured as the difference between the V_{DD} and the crossover point of Q and QB as shown in Figure 1.5.

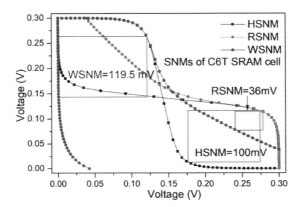

FIGURE 1.4 RSNM and WSNM values of conventional 6T SRAM observed at different subthreshold supply voltages.

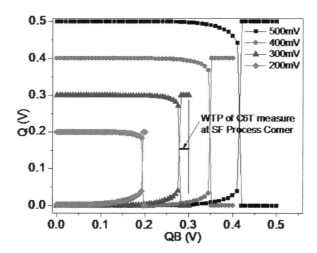

FIGURE 1.5 WTP of C6T SRAM measured at the crossover point of Q and QB at different V_{DD}.

1.4 DATA READ OPERATION IN SRAM

Figure 1.2 shows an SRAM cell being read. The bitlines are both initially precharged to high. Without loss of generality, assume Q is initially 1 and thus QB stored 0. When the WL is raised, bitline bar (*BLB*) should be pulled down through the driver and access transistors M3 and M5. Meanwhile, Q and *BL* both should remain at 1. At the same time, *BLB* must be pulled down, causing QB to rise. QB is kept low by M3 but is raised by the current flowing in from M5. Hence, driver M3 must be stronger than the access. Now, after WL assertion, the cell read current I_{CELL}, which is generated by the driver and access devices, causes a drop on one *BL*, which can be sensed by an SA with respect to the other to quickly decipher the accessed data. The timing waveform for the read operation is shown in Figure 1.6. Moreover, sense amplifiers play an important role in SRAM designs and will affect the speed and read failure rate for SRAM macros. SA sense the voltage difference at the *BL*s and generate a full-rail voltage swing at the outputs. In addition, sense amplifiers are used widely in digital circuits for a number of different important applications including SRAM, dynamic random access memory (DRAM) [106], input/output (I/O) [28], and Analog to Digital (A/D) conversion [76].

1.5 MISMATCH ISSUES IN SRAM AND FAILURE

In advanced CMOS technology nodes, the predominant yield loss comes from the increase in process variations, which strongly impact SRAM functionality as the supply voltage is reduced [75,62]. For the purposes of circuit design, the sources of variation can broadly be categorized into two classes that affect SRAMs:

1. Die to Die (D2D): also called global or inter-die variations, these affect all devices on the same chip in the same way (e.g., they may cause all the transistor's gate lengths to be larger than a nominal value).

Introduction

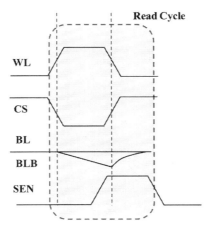

FIGURE 1.6 Read timing and voltage waveforms in the conventional SRAM.

2. Within Die (WID): also called local or intra-die variations, these correspond to variability within a single chip and may affect different devices differently on the same chip (e.g., devices in close proximity may have different V_{th} than the rest of the devices).

With this result, the impact of increased variations with the voltage scaling is more pronounced on the 6T SRAM cell. The rationed operations, both during read and write, leave the 6T bitcell highly susceptible to WID variation and manufacturing defects. The random threshold variations in sub-nanometer technologies have resulted in serious yield issues for realizing low-V_{DD} read/write operations with a 6T SRAM cell. Emphatically, SRAM scaling has become extremely difficult in the advanced technology nodes (e.g., 65-, 40-, or 32-nm low power (LP) CMOS technology). D2D variations have been a long-standing design issue and are typically accounted for during circuit design by using corner models [85]. These corners are chosen to account for the circuit behavior under the worst-possible variation and were considered efficient in older technologies where the major sources of variation were D2D variations. However, in nanometer technologies, WID variations have become significant and can no longer be ignored. The impact of the variation on the memory cell is elaborated in the following subsections.

1.5.1 READ FAILURE

During the read operation, the WL is activated for a short time determined by the cell read current, *BL* loading (capacitance), as shown in Figure 1.7. The content of a cell is read by sensing the voltage differential between the *BL*s using an SA. For a successful read operation, the *BL*s precharged to V_{DD} should discharge to a voltage differential value, which can trigger the SA correctly. Read failure occurs if the bitcell read current decreases below a certain limit, which often results from an increase in V_{th} for the pass gate (PG1), pull-down (PD1) transistors, or both. This decrease in I_{CELL} reduces the *BL* differential sensed by the sense amplifier, resulting

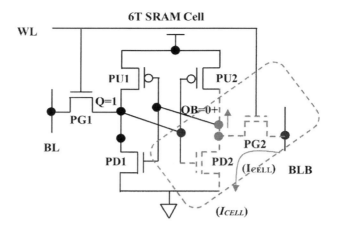

FIGURE 1.7 Schematic of 6T SRAM bitcell in the read operation.

in incorrect evaluation. This failure can also occur due to a large offset affecting the SA. This type of failure decreases memory speed because the WL activation time is about 30% of memory access time [75,62]. Similarly, from the SNM for read (RSNM) perspective, fast NMOS and slow PMOS results in 21% degradation in the cell stability.

1.5.2 Write Failure

The ability of the bitcell to be written correctly is referred to as write stability or write margin. In write operation, *BLB* is pulled to zero by the write driver, while WL is enabled, as shown in Figure 1.8. Therefore, the NMOS PG2 is turned ON/OFF, which results in a voltage drop in the storage node QB holding data 1 until it falls below $V_{DD} - V_{TH}$ for the PU1, where the positive feedback action begins. For a stable write operation, PG2 should be stronger than PU2. Due to WID variations, the pass gate cannot overcome the pull-up transistor, resulting in a write failure [85]. Write failure can also happen if the WL pulse is not long enough for the bitcell to flip the internal nodes (dynamic failure). Utilizing a high V_{th} cell decreases the write margin (WM) by 14%, and low VTH transistors result in a 36% degradation of cell stability. Regarding inter-die variations (considering process corners), slow NMOS (weak pass transistor) and fast PMOS (strong pull-up transistor) is the most difficult situation for write ability. This results in a 25% degradation in the WM compared to the nominal process corner [52].

1.5.3 Data Retention Voltage

In SRAM, the data retention voltage (DRV) defines the minimum V_{DD} under which the data in memory is still preserved. When V_{DD} is reduced to DRV, all six transistors

Introduction

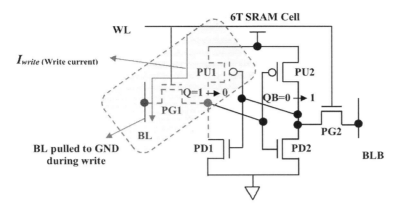

FIGURE 1.8 Schematic of 6T SRAM bitcell during the write operation.

in the SRAM cell operate in subthreshold region and therefore are strongly sensitive to variations [52]. DRV depends strongly on WID variations in the bitcell inverters, which may cause the bitcell to be imbalanced. If the bitcell is asymmetric due to WID variations, the bitcell tends to have a higher DRV than in the symmetric case. This can be explained using SNM, where the DRV voltage can be defined as the voltage when the hold SNM is equal to zero. In the symmetric case, both SNM high (upper left square) and SNM low (lower right square) decrease symmetrically to zero. However, in the case of the asymmetric bitcell shown in Figure 1.8, SNM low is always larger than SNM high, and the bitcell DRV is limited by the SNM-high case. Therefore, variations increase the bitcell DRV because they increase the asymmetry. Additionally, as scaling continues beyond the 90-nm node, the pushed rules used in bit cell design will warrant increased attention and more costly measures to avoid sources of systematic, nonrandom device mismatch. We define nonrandom mismatch as a mean offset in the device pair (e.g., pull-down NMOS left versus right) within the same or adjacent bit cell. Factors that may contribute to nonrandom mismatch are layout topology, process scaling practices, and use of pushed design rules in the bitcell.

1.5.4 Bitline Capacitance Mismatch

Difference between the true *BL* and complementary bitline *BLB* capacitive loads originating from the cell capacitor and metal interconnect becomes significant. To analyze this, let us consider a simple latch-based sense amplifier. In a n-latch pair, as shown in Figure 1.9a, when the latch is enabled, the currents through the load capacitances are

$$I_{BL} = C_{BL} \cdot \frac{dV_{BL}}{dt} = I_{DS,MN1} \text{ and } I_{BLB} = C_{BLB} \cdot \frac{dV_{BLB}}{dt} = I_{DS,MN2}.$$

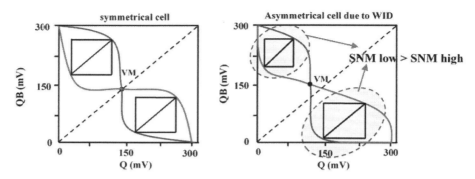

FIGURE 1.9 The voltage transfer characteristics (VTC) of (a) balanced and (b) imbalanced cells. VM is the trip point of the VTCs. The imbalanced bitcell has higher data retention voltage (DRV) than the balanced case because of asymmetry in the VTCs caused by variations.

Here $I_{DS,MN1}$ and $I_{DS,MN2}$ are currents of the saturated sensing transistors. If $C_{BL} = C_{BLB}$ and initial $V_{BL} = V_{BLB}$, the circuit is completely symmetric, and the voltages on both nodes change synchronously, that is, $\frac{dV_{BL}}{dt} = \frac{dV_{BLB}}{dt}$. However, once $C_{BL} \neq C_{BLB}$, the voltage changing speed of the two output nodes will be different, and the ground potential is inclined to appear on the node with smaller capacitance. Here, the initial voltage difference $V_{BL} - V_{BLB}$ gives an offset [84]:

$$V_{OS,n} = \frac{\sqrt{C_{BL}} - \sqrt{C_{BLB}}}{\sqrt{C_{BL}}} \cdot (V_{BL_{PRE}} - V_{thn}) \tag{1.8}$$

or,

$$V_{OS,n} \approx \frac{\Delta C}{2C} \cdot (V_{BL_{PRE}} - V_{thn}), \text{ when } C \gg \Delta C. \tag{1.9}$$

Equation 1.9 suggests the imbalanced bitline capacitance can be equivalent to an input offset voltage $V_{OS,n}$ that is applied to one input terminal of an n-latch sensing pair as shown in Figure 1.9b. With $V_{OS,n}$, the two output node voltages will drop at the same speed. Since $V_{OS,n}$ rises with the increase of the capacitance difference ΔC and equalization voltage V_{BLPRE}, high-level sensing is more sensitive to bitline capacitance mismatch. Therefore, bitline capacitance mismatch is very significant for the sense amplifier offset voltage. Any mismatch in bitline capacitance resulted in the killing of the differential voltage required for correct sensing, which increases the offset voltage of the sense amplifier. The bitline capacitance is a strong function of the physical design of the SRAM column and SRAM cell design. The sense amplifier may fail to read correctly due to asymmetry in an SRAM cell, which is a result of manufacturing process variations.

Introduction

1.6 RELIABILITY FAILURE MECHANISMS

Figure 1.10 depicts the different reliability mechanisms followed by a brief discussion on both time-zero (spatial) and time-dependent variabilities. Time-zero unreliability is independent of time and immediately visible right after production. Spatial unreliability effects are systematic (e.g., gradient effects, etc.) or random (e.g., line edge roughness [LER], random dopant fluctuation [RDF], etc.).

The effects depend on the circuit layout, the neighboring environment, process conditions, and the impact of the geometry and structure of the circuit and can lead to yield loss. This yield loss can be functional or parametric, that is, resulting in malfunctioning circuits or circuits with degraded performance, respectively.

Temporal unreliability effects, by comparison, are time-varying and change depending on operating conditions such as the operating voltage, temperature, switching activity, and the presence and activity of neighboring circuits. A difference is made between wear-out or aging effects (e.g., hot carrier injection [HCI], negative bias temperature instability [NBTI], etc.) and transient effects (e.g., electromagnetic interference [EMI], single-event upsets [SEUs], etc.). In the following subsections, these effects are discussed in more detail.

1.6.1 Time-Zero Variability

Spatial unreliability or process variability is an increasing problem in nanometer CMOS IC production. The problem results from the increasing complexity needed to fabricate nanometer CMOS devices, combined with the scaling toward atomistic device dimensions (<180-nm CMOS). Typically, the parametric yield is used as a metric to express the impact of these effects on the performance of the circuit right after production. A high yield implies low spatial unreliability.

Spatial unreliability or time zero ($t = 0$) variation occurs during device fabrication at the start of the chip's lifetime. These variations are due to an imperfect device

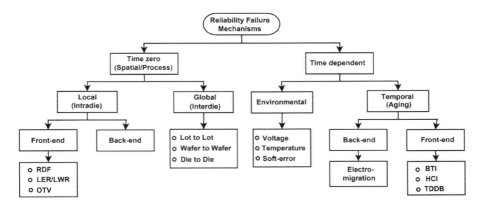

FIGURE 1.10 Classification of reliability failure mechanisms [49].

fabrication process [99,94]. The device parameters are deviated from their expected value during the manufacturing period because of limited controllability in the fabrication process at deep submicron technology [67]. Process variation is a phenomenon of variation in the device parameters such as channel length (L), channel width (W), and gate oxide thickness (t_{ox}). Increased variability in device parameters due to process variation decreases total yield, which creates a barrier to future scaling. Due to further technology scaling the focus has shifted to process variation. There are two types of time-zero variability:

1. Inter-die (global) variation
2. Intra-die (local) variation

1.6.1.1 Inter-Die/Global Variation

The inter-die variation is the parameter variations across dies that emerge from various wafers and lots as depicted in Figure 1.11 [66]. The inter-die variation is caused by the deviation in the photolithographic process and parameter variation on the same device manufactured at varying times over a larger area. It is systematic in nature and affects all devices equally in a die. It leads to five different process corners, namely, fast–fast (FF), fast–slow (FS), slow–fast (SF), slow–slow (SS), and typical–typical (TT) as shown in Figure 1.12. FF, SS, and TT are uniform corners as they uniformly impact both PMOS and NMOS devices, whereas SF and FS are nonuniform process corners as they affect the NMOS and PMOS devices unevenly. Typically, the circuits that are designed for fixed V_{DD} and manufactured

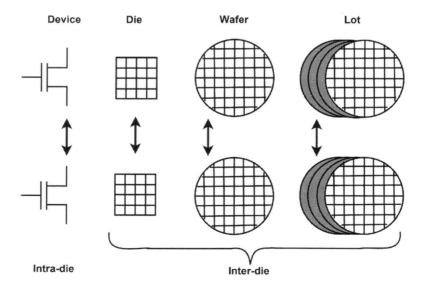

FIGURE 1.11 Process variations at different phases of fabrication.

Introduction

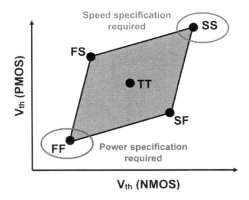

FIGURE 1.12 The SPICE box of different process corners.

with the FF corner have the highest frequency, provide the best performance, and have maximum leakage power, which required power optimization schemes. Similarly, circuits manufactured in the SS corner will have the slowest performance, which requires speed specification; have the least leakage; and consume less power.

For circuit design, inter-die variation is accounted as a shift in the mean (μ) of some electrical parameter (e.g., threshold voltage) equally across all devices [66]. Beyond the 45nm technology node, the inter-die variability is generally larger than intra-die. This variation across the die may impact the performance and parametric yield.

1.6.1.2 Intra-Die/Local Variation

In contrast to the systematic inter-die variation, this variation is due to parameter mismatches across the two identical devices placed next to each other on the same die [68]. The intra-die variation is unpredictable in nature and is caused by random uncertainties in the fabrication process, such as variations in oxide thickness, length, width, number of dopants, flat band control, and the like [24]. Figure 1.10 depicts the local variations divided into the front end and the back end. The front end comprises crucial intrinsic parameter fluctuations, such as line edge/width roughness (LER/LWR) [109], RDF [61], and oxide thickness variations (OTVs) [47]. The intra-die variation strongly depends on the circuit layout and exhibits spatial correlation, that is, devices that are close together in the layout have a higher probability of being alike in characteristics than devices placed far apart. As the device dimensions scale down into the nanometer regime, intra-die variation becomes critical for circuits. It is also an important parameter for analyzing circuit performance and predicting the yield of a chip.

In addition to time-zero variation, there is another type of variation at a time greater than zero, and it is known as time-dependent variation. This type of variation is explained next.

1.6.2 Time-Dependent Variability

With the continuous technology scaling in the past few decades, the time-dependent (t = 0) aging degradation becomes the major contributor to device unreliability. There are various time-dependent failure mechanisms, which affect the reliability of CMOS devices/circuits. Environmental and temporal (aging) variations are the two sources of time-dependent variability. The following sections briefly describe these time-dependent variability issues.

1.6.2.1 Aging/Temporal Variations

Aging phenomena in ICs were first observed four decades ago. At that time, researchers were mainly focused on understanding these effects rather than solving circuit reliability problems. Later, due to the aggressive scaling of the device dimensions and the increasing electric fields, circuit aging becomes a severe issue. After the turn of the century, to further scale CMOS technologies, the introduction of new materials introduced additional failure mechanisms and made existing aging effects more severe [57].

Among many reliability challenges ranging from an increasing number of design defects to a higher sensitivity to intermittent and transient failures [40,39], the accelerated temporal degradation of CMOS devices/circuits is the major concern for recent-generation ICs. Based on the environmental conditions and usage of ICs, aging is the failure mechanism that slowly widens its initial distribution [72].

Temporal degradation during different operating phases in CMOS inverter is shown in Figure 1.13. The major time-dependent temporal reliability concern at deep submicron technologies are the following:

HCI
Time-dependent dielectric breakdown (TDDB)
Bias temperature instability (BTI)
Electromigration (EM)

HCI occurs during the switching of logic, whereas BTI and TDDB occur during static logic output. The BTI is considered the most important failure mechanism in the research community, and it is the main focus of this work.

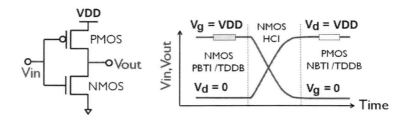

FIGURE 1.13 Temporal degradation during different operating phases in a CMOS inverter.

Introduction

1.6.2.2 Environmental/Transient Variations

Transient or environmental variations vary with the operation and lifetime of the circuit. Therefore, it is important to analyze the effect of time-dependent environmental variations on the circuit and implement the design techniques to prevent them. Temperature variations, voltage variations, and soft errors are major time-dependent environmental variations.

1.6.2.2.1 Supply Voltage Variation

The variation in supply voltage affects the operating speed of the MOSFETs. The supply voltage fluctuations are mainly caused by IR drop and di/dt noise. IR drop is caused by the current flow over the parasitic resistance of the power grid, whereas di/dt noise is due to the parasitic inductance in combination with capacitance and resistance of the power grid and package. These fast-changing effects (also called power noise) typically have time constants in the range of nano- to microseconds [102].

The fluctuation in switching activity of the circuit causes a nonuniform power requirement and may lead to logic failures. The subthreshold leakage variations of the transistor also impact the supply voltage distribution across the circuits. The change in supply voltage affects the performance of the circuit and failure rates due to variability. The delay of the CMOS logic gate can be approximated by [79]

$$t_{gate} \propto \frac{V_{DD}}{b(V_{DD} - V_{th})^a}, \qquad (1.10)$$

where a and b are the gate-specific fitting parameters. The reduction in supply voltage increases the delay of the circuit, which leads to a degradation in performance, whereas increasing supply voltage compensates for the performance and significantly reduces the circuit failure rates due to variability [80].

1.6.2.2.2 Temperature Variation

The temperature of the chip is affected by the power dissipation, which depends on the thermal conductivity. The power dissipation leads to both local and global temperature variations. The local temperature fluctuations in high-activity regions are called hot spots, whereas ambient temperature variation globally shifts the chip temperature.

Temperature variations affect the performance and operating conditions of the MOSFET. With increasing temperature, the threshold voltage of the MOSFET reduces, which positively impacts the delay and decreases the carrier mobility, which negatively affects the delay and consequently increases the leakage current. The temperature dependence on threshold voltage is given by [98]

$$V_{th} = 2\phi_F - \frac{Q_{ss}}{C_{ox}} + \phi_{ms}, \qquad (1.11)$$

where ϕ_F is the Fermi potential, Q_{ss} is the surface charge at the Si–SiO$_2$ interface, C_{ox} is the gate oxide capacitance, and ϕ_{ms} is the work function difference and is the function of the temperature [98].

$$\phi_{ms} = -0.61 - \phi_F(T) \tag{1.12}$$

Here, $\phi_F(T)$ is the temperature-dependent Fermi potential. Equation 1.12 shows that the work function difference reduces with an increase in temperature and hence decreases the threshold voltage.

1.6.3 Bias Temperature Instability

Bias temperature instability (BTI) is time-dependent and is caused due to dangling bond defects at the Si/SiO$_2$ interface that allows the trapping of charges even at small energies into the Si/SiO$_2$ interface, which increases the threshold voltage and affects both the NMOS and PMOS devices [71]. NBTIs and positive BTIs (PBTIs) are observed in both PMOS and NMOS. However, the effect of NBTIs in PMOS is more sizable and is the dominant limiting factor of a device's/circuit's lifetime comparable to all other components (NBTIs in NMOS, PBTIs in NMOS, PBTIs in PMOS); hence, only NBTIs in PMOS is usually considered [100,81]. The effect of PBTIs is expected to increase with the acceptation of high-k dielectrics in the gate oxide to reduce the leakage current [12]. NBTIs leads to significant shifts in the V_{th} of PMOS over time, which creates uncertainty in device/circuit behavior and decreases the lifetime of the device/circuit. Because of this progressive degradation, it is difficult to ensure the reliability of the IC over its lifetime [87]; hence, it is essential to explore variation-tolerant design solutions to mitigate variability issues. Figure 1.14 shows that the lifetime of an inverter is degraded due to NBTI by 2.2× for every 10°C increase in temperature, which means the lifetime of any circuit depends on the operating temperature [26].

1.6.3.1 BTI Models
The two well-known BTI models are reaction–diffusion (RD) model and the trapping–detrapping (TD) model [90]. The RD model is one of earlier developed models for analyzing the impact of aging in metal-oxide semiconductor (MOS) devices [82,7,56].

1.6.3.1.1 RD Model
The RD model is based on the concept of the breaking of the silicon–hydrogen bond at the SiO$_2$ interface of the NMOS or PMOS transistor [31]. As a result, the RD model has two phases consisting of a stress phase and a recovery phase as shown in Figure 1.15 [38,45]. At the stress phase, the silicon–hydrogen bond is broken at the gate of either an NMOS or a PMOS device. The broken silicon bond stays at the interface (traps or charges) while either the hydrogen or its molecule diffuses toward the device gate. However, during the recovery phase, there is no breaking of the silicon–hydrogen bond; as a result, the hydrogen or its molecule diffuses back to the interface, which is also known as recovery or annealing.

Introduction

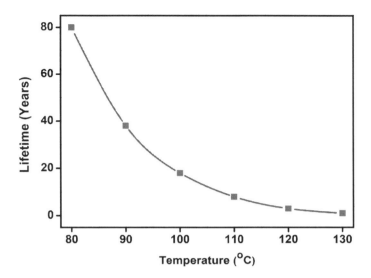

FIGURE 1.14 Lifetime degradation of an inverter due to NBTI under temperature variation [26].

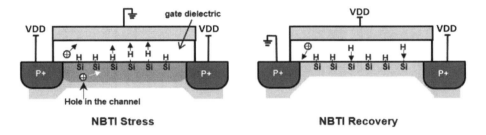

FIGURE 1.15 The illustration of NBTI stress and recovery.

There are two different types of NBTI:

1. **Static NBTI:** When the transistor is under constant stress without recovery.
2. **Dynamic NBTI:** When both the stress and recovery phases alternatively exist in the transistors. The effects of dynamic stress on circuit timing error are illustrated in Figure 1.16.

The effect of NBTI can be modeled for both the cases of stress in terms of threshold voltage shift (ΔV_{th}).

The static NBTI is expressed as

$$\Delta V_{th}^{\text{static}} = A\left((1+\delta)t_{ox} + \sqrt{C(t-t_0)}\right)^{2n} \tag{1.13}$$

$$A = \left(\frac{qt_{ox}}{C_{ox}}\right)\sqrt[3]{K^2 C_{ox}(V_{gs}-V_{th})e^{\left(\frac{E_{ox}}{E_0}\right)^2}}, \tag{1.14}$$

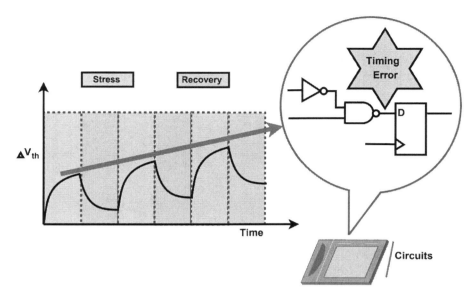

FIGURE 1.16 Effect of dynamic stress on circuit timing error.

where q is the electron charge, E_{ox} is the gate oxide electric field, C_{ox} is the oxide capacitance per area, and n is the time-dependent factor that lies between $\frac{1}{4}$ to $\frac{1}{6}$. All other constants are given in Table 1.3.

The expression of stress and recovery phase for dynamic NBTI is given as

$$Stress: \Delta V_{th} = \left(K_v (t-t_0)^{1/2} + \sqrt[2n]{\Delta V_{th0}} \right)^{2n} \quad (1.15)$$

$$Recovery: \Delta V_{th} = \Delta V_{th0} \left(1 - \frac{2\xi_1 t_e + \sqrt{\xi_2 C(t-t_0)}}{2t_{ox} + \sqrt{Ct}} \right), \quad (1.16)$$

where all the coefficients and constants are given in Table 1.3. These equations are used to model the long-term NBTI effect when the transistor undergoes alternative to stress and recovery phase for a long time.

$$\Delta V_{th}^{dynamic} = \left(\frac{\sqrt{K_v^2 \beta T_{clk}}}{1 - \beta_t^{1/2n}} \right)^{2n} \quad (1.17)$$

$$\beta_t = 1 - \frac{2\xi_1 t_e + \sqrt{\xi_2 C(1-\beta)T_{clk}}}{2t_{ox} + \sqrt{Ct}}, \quad (1.18)$$

where T_{clk} is the clock cycle and β is the duty cycle. According to the measurement performed by different industry terms on various technology processes, NBTI-induced ΔV_{th} is a strong function of the duty cycle [55].

Introduction

TABLE 1.3
RD Model of NBTI-Induced ΔV_{th}

NBTI-induced ΔV_{th}		
Static		$A\left((1+\delta)t_{ox} + \sqrt{C(t-t_0)}\right)^{2n}$
Dynamic	Stress:	$\left(K_v(t-t_0)^{1/2} + 2\sqrt[n]{\Delta V_{th0}}\right)^{2n}$
	Recovery:	$\Delta V_{th0}\left(1 - \dfrac{2\xi_1 t_e + \sqrt{\xi_2 C(t-t_0)}}{2t_{ox} + \sqrt{Ct}}\right)$
	Long-term:	$\left(\dfrac{\sqrt{K_v^2 \beta T_{clk}}}{1-\beta_t^{1/2n}}\right)^{2n}$

Constants and coefficients	
A	$\left(\dfrac{qt_{ox}}{C_{ox}}\right)^3 \sqrt{K^2 C_{ox}(V_{gs}-V_{th})} \left(e^{\frac{E_{ox}}{E_0}}\right)^2$
K_v	$\left(\dfrac{qt_{ox}}{C_{ox}}\right)^3 K^2 C_{ox}(V_{gs}-V_{th})\sqrt{C}\left(e^{\frac{E_{ox}}{E_0}}\right)$
E_{ox}	$\dfrac{V_{gs}-V_{th}}{t_{ox}}$
C	$T_0^{-1} \exp(-E_a/KT)$
t_e	t_{ox} if $t-t_0 > t_1$
	$t_{ox}\sqrt{\dfrac{t-t_0}{t_1} - \sqrt{\dfrac{\xi_2 C(t-t_0)}{2\xi_1}}}$; otherwise
t_0	time at which the stress phase begins
t_1	time at which the recovery phase begins
E_a (eV)	0.49
E_0 (V/nm)	0.335
δ	0.5
K ($s^{-0.25}C^{-0.5}nm^{-2}$)	8×10^4
ξ_1	0.9
ξ_2	0.5
T_0	10^{-8}

Although the RD model has been mostly used by several researchers, there are still questions about whether it is accurate enough as an ideal or near-ideal aging model. Nevertheless, TD model was developed as an alternative and accurate aging model.

1.6.3.1.2 TD Model

According to the TD model, the BTI of each device is characterized by three different factors [34,35]:

1. Number of defects (n)
2. Defects capture time (τ_c): The time needed to change a gate oxide defect during the stress phase
3. Defects emission time (τ_e): The time needed for the defect to reemit its charge during the recovery phase

The total number of defects is modeled based on a Poisson distribution:

$$n = Poiss(N_T), \tag{1.19}$$

where N_T is the occupied traps. The effects of each occupied trap on the threshold voltage of the transistor are obtained from an exponential distribution:

$$\Delta V_{thi} = Exp(\eta), \tag{1.20}$$

where η is the average impact of the individual traps on the threshold voltage $\left(\propto \dfrac{1}{device\ area}\right)$. The cumulative distribution function of the total BTI-induced ΔV_{th} is obtained by [34]

$$H_{\eta,N_T}(\Delta V_{th}) = \sum_{n=0}^{\infty} \frac{e^{-N_T} N_T^n}{n!}\left[1 - \frac{n}{n!}\Gamma\left(n, \frac{\Delta V_{th}}{\eta}\right)\right]. \tag{1.21}$$

The η value is extracted from experiments [23]. The N_T value is calculated using capture/emission time (CET) maps.

1.7 SOFT ERRORS

Aggressive scaling of feature size has additionally made transistors more vulnerable to radiation-induced single-event effects (SEEs) due to decreased nodal capacitances and supply voltages. The most common SEE of a digital device is the SEU, that is, flipping of logic states induced by the particles. This problem has begun affecting SRAMs even in space and terrestrial environments. In space and terrestrial environments, neutron radiation (1–1000 MeV) and alpha particles from cosmic rays affect the SRAM performance. These alpha particles from cosmic rays are discharged by radioactive impurities inside the packaging substances of chips [27,8].

Introduction 25

ICs' sensitivity to SEUs is directly related to the electrical characteristics of the transistor. Thus, drift in parameters due to a degradation mechanism like NBTI may induce variations in the SEU sensitivity of the device throughout its lifetime [19]. The new technologies convey our consideration regarding neutrons, which, by implication, induce soft errors through a response to the core of transistor materials as displayed in Figure 1.17. The charged secondary particles like alpha particles, protons, and heavy ions are produced by the nuclear reaction, and they create electron–hole pairs with the aid of direct ionization on the particle track and store charge. The produced charge is collected to drain by drift and diffusion and causes a soft error [93,104].

Along with susceptibility to NBTI, aggressive scaled-down CMOS devices are more sensitive to radiation-induced errors [9]. Radiation-influenced SEEs are mainly due to reduced supply voltages and node capacitances. An SEU is the substantially common SEE in storage elements (latches and memory cells), which flip the circuit logic states. In terrestrial and space environments, alpha particles and neutron radiation from cosmic rays affect circuit performance [27,8]. Moreover, in recent CMOS technology, it has been proven that alpha-particle generation induced by neutron hits greatly contributes to the overall soft error rate in terrestrial and space environments [4]. Direct ionization from secondary protons is a visible, undeniable reality that a CMOS circuit low-voltage operation reduces the critical charge (Q_{crit}) [4,83]. Additionally, as Chandra and Aitken [14] have shown, the decrease of critical charge with supply voltage reduction has been studied in detail. The critical charge is the minimum amount of collected charge at the sensitive node of the circuit during particle strike which is sufficient to make equal input and output voltages of the circuit. On top of that, NBTI exacerbates soft-error-rate (SER), since it causes Q_{crit} reduction with the stress time [77].

In current technology, it is accounted for by direct ionization because the secondary alpha particles are noteworthy sponsors to the neutron soft-error rate in space and terrestrial environments [4]. The direct ionization from secondary protons can be a noteworthy patron in light of the fact that the low-voltage operation of SRAM

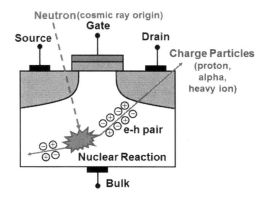

FIGURE 1.17 Soft-error mechanism in semiconductor devices by a neutron.

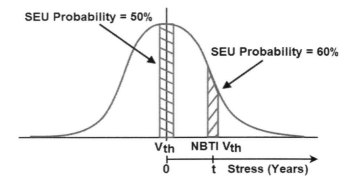

FIGURE 1.18 SEU probabilities for V_{th} bin and the NBTI-shifted V_{th} bin [37].

decreases the Q_{crit} [4,83]. The NBTI also impacts considerably the Q_{crit} of the circuit node and Q_{crit} reduces with the stress time [77].

Impact of NBTI Aging on SEUs

Figure 1.18 demonstrates the comparative SEU probabilities for the normal threshold voltage and the NBTI-caused threshold voltage in the framework of the V_{th} parameter range [37]. The proportion of the NBTI-caused SEU probability to the normal SEU probability represents the average increase in SEU likelihood. From the figure, we can observe that the effect of SEU increases with an increase in stress time. High radiation, fault tolerance, and the mitigation of the NBTI effect can be achieved by the circuit-level designs [37,25].

1.8 FIN-SHAPED FIELD-EFFECT TRANSISTORS AND ALTERNATIVE TECHNOLOGIES FOR RELIABLE SRAM DESIGN

Scaling has directly or indirectly been the root cause of the tremendous capabilities of today's ICs and their ubiquitous use in nearly all modern electronic systems. As the silicon industry is moving toward the end of the technology road map, controlling the fabrication of scaled MOS devices is becoming a great challenge [42]. Another important challenge for the MOS devices is the diminishing gate control over the channel, which manifests itself in the form of increased short-channel effects (SCEs) and leakage currents. Furthermore, increasing inter-die/intra-die statistical variations in the process parameters, such as channel length (L), width (W), and transistor threshold voltage emerged as a serious problem in the nanoscaled circuit design [21].

1.8.1 FIN-SHAPED FIELD-EFFECT TRANSISTORS

Various device structures have been proposed so far to mitigate the severe problem of SCEs and variability (Figure 1.19). However, due to its superior gate control,

Introduction

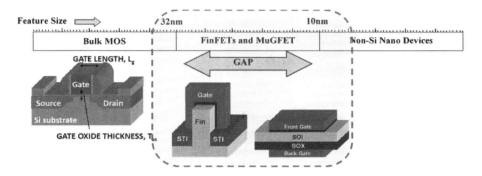

FIGURE 1.19 Various device structures to mitigate severe problems of SCEs and variability.

improved electrostatic integrity, and variability, the fin-shaped field-effect transistor (FinFET) has demonstrated satisfactory scalability and feasibility for the mass production of post-22-nm nodes. Therefore, FinFET is an ideal alternative to planar CMOS in SRAM designs. In order to effectively suppress SCEs, the thickness of the body (i.e., the fin width) should be no greater than 2/3× the gate length (L_G) [15]. Although significant circuit solutions are available for performance and reliability enhancement of sense amplifiers, however, increased inter-die/intra-die variation in MOSFET technology in submicron technology makes the design space very limited for the circuit designer and researchers. In view of this and to align with the latest industry standards, we explore the sense amplifier design with multigate MOSFETs, specifically the double-gate FinFET.

The following salient feature of FinFET over MOSFET makes it a promising candidate for SRAM design. The promising matching behavior makes it more immune to variability, resulting in improved offset characteristics.

1. Excellent control of short channel effects in submicron regime and making transistors still scalable. Due to this reason, the small-length transistor can have a larger intrinsic gain compared to the bulk counterpart.
2. Much lower off-state current compared to bulk counterpart
3. Promising matching behavior
4. Fully depleted structure, enabling better on/off contrast
5. Faster switching speed (it comes from lower input capacitance and higher dynamic current density)

In addition, SEU-sensitive volume is also decreased in the FinFET structure in the SRAM cell design and can offer elevated resistance against SEUs compared to the bulk CMOS counterparts [20,41].

1.8.2 Silicon on Insulator

The main difference between conventional MOS structure and silicon on insulator (SOI) MOS structure is that an SOI device has a buried oxide layer, which

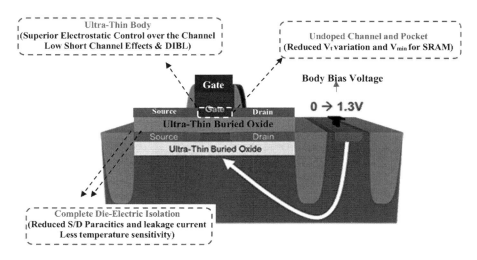

FIGURE 1.20 SOI device.

isolates the body from the substrate. As shown in Figure 1.20, the SOI transistor is a planner device. In conventional MOS technology, with the shrinking of channel length, the source and drain come into close proximity, and the gate terminal loses control over the channel, which is highly undesirable. Consequently, the impact of this is to increase the subthreshold leakage current from drain to source terminal. This results in more power consumption and incurs power-hungry memory designs. This is also a severe bottleneck in scaling the MOS devices and SRAM footprint in the SOC. In addition, enhanced variability due to inter-die/intra-die variation is a major challenge for SRAM scaling below sub-22-nm nodes, this gives an opportunity to explore alternate technologies for energy-efficient SRAM design. In particular, the impact of this process variations on device threshold voltage (V_{th}) due to RDF and LER increases significantly with the channel length (LG) are scaled down to below 28 nm [107]. This increased transistor mismatch results in a higher variation in SRAM cell design margins and read/write failures.

The technique to reduce RDF effects, the body should be undoped [107]. In FinFETs, LER is a severe bottleneck thanks to the spacer lithography techniques, which can be mitigated by this [91]. The SOI offers improved electrostatic integrity than the bulk structure and has a process flow and layout with additional contact, like that of the conventional MOSFET [32]. Fully depleted SOI (FD-SOI) have an ultra-thin layer of an insulator called the buried oxide. Initially, the buried oxide (B_{ox}) is deposited on top of the silicon. Later, a very thin silicon film is epitax to form a transistor channel. There is no need to do doping in the channel, because of the thin film structure, emphatically making the transistor fully depleted.

The device threshold in the FD-SOI technology is vigorously dependent on the body thickness (T_{Si}). However, advancement in the process technology enables the FD-SOI with very regular wafer thickness and variation below 0.5nm, this enables

TABLE 1.4
SOI Device Parameters Transistor

Parameters	NMOS	PMOS
LG (nm)	25	25
$L_{\it eff}$ (nm)	35.6	30.7
t_{ox} (nm)	1	1
T_{box} (nm)	10	10
T_{si} (nm)	6	6
ϕ_m (eV)	4.45	4.85

scaling of channel length to well beyond 22nm. In order to reduce RDF-induced variations, in the ultra-thin body, FD-SOI devices are fabricated via implantation-free process to reduce dopant straggling and damage-induced defects [48].

1.8.3 Tunnel Field-Effect Transistors

Subthreshold leakage current is a very severe bottleneck in deep submicron technology node MOSFETs. Since SRAM occupies a significant area (>50%) in SOCs, leakage power dissipation is a major issue in achieving low-power chips [10]. Furthermore, the exponential increase in on-chip transistor count has also increased the overall power consumption, resulting in performance/watt of energy consumption, a very important design constraint in current high-performance SRAMs. Aggressive V_{DD} scaling while maintaining the device performance is a vital technique for reducing energy consumption. This approach reduces the dynamic power quadratically and the leakage power consumption linearly. However, enabling V_{DD} scaling by reducing the transistor threshold voltage increases the subthreshold current exponentially as manifested in Equation 1.22 [97]:

$$I_{off} = I_{DS} \times 10^{\left(\frac{-V_T}{SS}\right)}, \qquad (1.22)$$

where $SS = \eta\left(\dfrac{kT}{q}\right) \ln 10$ and $\eta = 1 + \left(\dfrac{C_{dep}}{C_{ox}}\right)$. In Equation 1.22, SS is the subthreshold slope of the device, and it depends on depletion capacitance, C_{dep}, and oxide capacitance; C_{ox}, V_T is the thermal voltage. Furthermore, ultra-thin gate oxide causes a significant increase in the gate current of the transistors. This causes huge off-state power consumption in SRAMs. Steep subthreshold devices are promising contenders to replace the conventional MOS devices for subthreshold leakage. The tunnel field-effect transistor (TFET), shown in Figure 1.21, is a promising candidate and operates at a very low threshold voltage with low leakage and scaled V_{DD}. TFET works on the principle of inter-band tunneling of carriers from the valence band in the source to the conduction band in the channel [33]. In contrast to conventional

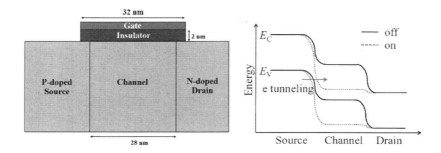

FIGURE 1.21 TFET device structure and energy band diagram.

MOSFETs, TFET is not a symmetric device, source/drain terminals are not interchangeable and must be defined at the time of fabrication. Consequently, TFETs conduct current in unidirectional only, and pose a critical challenge in the design of access transistors in 6T SRAM cell, since access transistors are required to conduct current in both directions.

SRAM read and write latency is a very important figure in its merits; however, TFET-based SRAMs suffer from low read/write time due to low on current (I_{on}) in TFET devices. To cope up with this issue, GaSb/InAs heterojunction TFET (Hetro-j TFET) is a very promising technique due to its broken-gap band alignment, resulting in higher drive current [63,96]. To overcome the effect of short channel effects in TFET-based SRAMs, source underlapped GaSbInAs-based vertical double-gate engineered nTFET/pTFET are explored and offer a significant advantage in sub-10-nm technology nodes. However, the ion for both underlapped and overlapped structures is the same (210 µA/µm), and less-than-conventional bulk MOSFETs give no advantage for SRAM read/write latency.

1.9 APPLICATIONS OF ULP SRAM

The conventional 6T and 8T SRAM architectures proposed earlier have benefited memory design industries, which have targeted designing high-speed, high-performance cache memory for high-end processors for computer applications. However, with development in new-generation bulk CMOS technologies and the requirement of ULP SRAMs for applications in handheld electronic devices, that is, mobiles, laptops, field programmable gate arrays (FPGAs), Internet of Thing (IoT) edge devices, and image/video-processing systems, these conventional SRAMs fail to achieve the trade-off between system power and performance. Thus, the development of application-oriented SRAM architectures working at subthreshold voltages has emerged significantly. Accordingly, the objectives of this book are to develop SRAM topologies targeted toward applications in FPGA, IoT edge devices, and image processing for object detection and tracking systems. Moreover, the application of the subthreshold SRAMs is explained in more detail in the succeeding subsections.

1.9.1 APPLICATION OF SRAM IN LOOK-UP TABLE OF FPGA

Digital devices use FPGAs as a platform for implementing digital circuits and systems [86]. Moreover, cache memory and FPGA devices use SRAM cells as bit storage devices. Since customers demand a longer power backup for their portable devices, research on power reduction in SRAMs and FPGAs has emerged significantly. FPGAs offer a short time to market and low design cost, which makes them gradually more prominent in the present market. However, due to their design flexibility, FPGAs do not attain similar area, delay, and power compared to application-specific integrated circuits (ASICs) [103]. This is mainly due to the overhead logics used for the reconfigurability. Considering FPGA devices as a platform for implementing a digital system at ultra-low-voltage (ULV) operations, it is desirable to focus on the performance of internal logic blocks of FPGA. The basic logic element (BLE) and look-up table (LUT) are the basic building blocks of FPGA as shown in Figure 1.22a. It is observed from Figure 1.22a that the LUTs occupy a huge amount of area in an FPGA. Therefore, to reduce the overall power of an FPGA, a reduction in LUT power is essential. Moreover, the FPGA device uses an SRAM cell as a memory element in the LUT as shown in Figure 1.22b. SRAM-based FPGAs designed by Xilinx and Altera cover the major portion of the global market share. These FPGAs use SRAM bitcells for routing and programmable computational functions, typically through LUTs and multiplexers as shown in Figure 1.22b. An FPGA has thousands of slices, and each slice has BLEs, which consist of numerous LUTs. In a single six-input LUT, there are 64 SRAM cells present to store and carry the store values to the output using the 6-input lines.

However, it is noted from the preceding analysis that to control the overall power of an FPGA, the focus should be on a power reduction in the SRAM [18,16,53]. The two main components of an FPGA, which depend on programmable elements (PEs) are the BLEs and the routing switches that connect the BLEs. The performance, power consumption, and functionality of these components depend on the PEs used. A typical logical element comprises a 6-input LUT, multiplexers, and flip-flops. Figure 1.22b provides examples of a generic logic element with a 6-input LUT that can be used to implement any 6-input function. The LE, LUT, and the multiplexer need configuration data that will control the functionality of the logic block. For example, the 6-input LUT requires 64 look-up values, and the multiplexer requires the information about whether to select the '1' input or the '0' input. The control logic 'M' characterizes whether the digital system is sequential or combinational (sequential for M = 1 and combinational for M = 0). This configuration data must be supplied to the linked PEs before the logical elements can be used. However, SRAM-based FPGA devices are designed at different technology nodes by numerous FPGA firms [60,59,22,5,17].

In addition, the ULP SRAM design is essential in embedded systems such as biomedical sensing devices, battery-operated wireless sensors, and portable electronic devices in which battery life and input power are the key concerns. The best way to reduce the power consumption in a digital IC is to lower the supply voltage as it has a direct effect on the power consumption. SRAM design

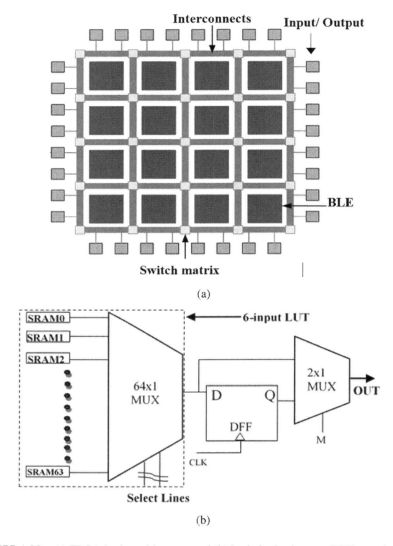

FIGURE 1.22 (a) FPGA logic architecture and (b) basic logic element (BLE) consists of a 6-input LUT, a D flip-flop (D-FF), and a 21 multiplexer [P10]-[P17].

remains challenging and is becoming more exciting due to the rapid development in CMOS technologies and the increased demand for on-chip memory in wireless implantable and wearable biomedical sensors. Subsequently, SRAM memory contributes as the major source of power consumption in an electronic device due to the introduction of leakage in the subthreshold region. Thus, power consumption in SRAM-based FPGAs and cache memories need to be revived in ULV operations.

1.9.2 Application of SRAM in IoT Edge Devices

The high standby power consumption in IoT devices has directed researchers and scientists toward the growth of ULP SoCs that are capable of functioning at sub-threshold voltages [46,108]. One of the ways to obtain ULP and energy-efficient IoT devices is by scaling down the supply voltage [11]. Though, due to low supply voltage the on-to-off current ratio (I_{ON}/I_{OFF}) also reduces. In addition, the exponential dependency of the current on the threshold voltage in the subthreshold operating voltages presents numerous challenges, particularly in rationed circuits, that is, a 6T SRAM bitcell. Subsequently, conventional SRAMs consume higher read and write power; therefore, scaling supply voltages and methods for reducing leakage at subthreshold voltages are required. Nonetheless, the low read and write stability is also the foremost issue, which resists conventional SRAM architectures to work in the subthreshold region [73].

Due to the swift progress of online markets, IoT carries connectivity, communication, and data collection to the current devices. The IoT brings countless devices connected and interconnected with each other to improve the present lifestyle as shown in Table 1.5. The application of IoT ranges broadly from the old-style internet to the industrial internet to the consumer internet [74]. The SoC block of IoT as shown in Figure 1.23 encompasses numerous subsystems like communication ports, high-end processors, sensors, security blocks, and on-chip memories, which comprise read-only memory (ROM) and SRAM. However, the SRAM has been divided into two types of systems, namely, fast and low power. Each of them has its own merits, features, and applications. The IoT devices, where SRAMs are used as memory storage, need it be for either high speed or low power consumption depending on the application. Yet there is a very high demand for high-performance devices with ULP consumption to implement complex operations while running on portable devices. This demand is typically driven by new-generation medical devices, handheld devices, and communication systems, which are fulfilled by IoT devices. The expansion of IoT is led in two distinctive directions, that is, smart wearables and automation. The wearable devices use SRAM as a memory element, which has a small footprint and low power consumption. Thus, SRAMs with high speed, high stability, and low power consumption offer substantial importance to IoT devices and their applications.

Moreover, the IoT portable devices communicate with each other and thus require a huge amount of memory to store and process data. The memory requirement in IoT depends on the applications related to the market. For instance, in the case of a huge amount of data storage and handling information for a long time, a

TABLE 1.5
Internet of Things (IoT) Devices and Their Applications

IoT devices	Applications
IoT Wi-Fi Devices (LAN)	Instruments, health care, entertainment, appliances
IoT Short-Range Devices (PAN)	Peripherals, drones, toys, remotes, wearables
IoT Long-Range Devices (WAN)	Automotive, industrial, security, infrastructure

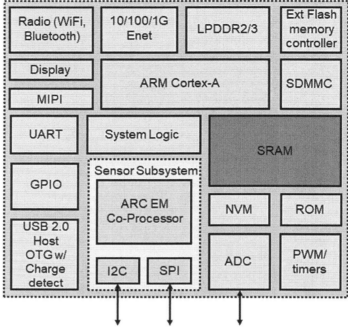
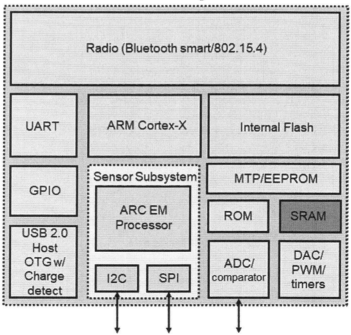

FIGURE 1.23 Alternative architectures for IoT devices [3].

Introduction   35

low-power high-density memory is required. By comparison, for high data transfer rate systems, a fast SRAM memory is required, whereas high speed is essential to communicate between IoT devices. Therefore, SRAM is always preferred as a cache memory due to its faster response. Moreover, the robustness of such memory systems irrespective of the variations in process–voltage–temperature (PVT) values of MOS devices and power efficiency are two of the most important design constraints [73]. As per the literature, more than 40% of the active energy is consumed due to the leakage current in modern high-performance processors [51,78]. Moreover, most of the time, a large number of SRAM cells used in present-generation IoT on-chip cache memory stay in a hold state during which leakage power dominates over the dynamic power. Thus, leakage reduction has become an imperative concern in the SRAM memory design in IoT edge devices. High-end IoT edge SoC blocks in cell phone and handheld device architectures typically contain different cores of the processor with memory management units (MMUs), external DRAM, liquid crystal display (LCD) controllers, and graphics processing units (GPUs) as shown in Figure 1.23.

The memory size required for high-processing IoT edge devices is more than compared to low-end IoT edge devices. The IoT processors in high-end devices require fast read–write access with high stability and low processing power. Therefore, they are currently built on 22-nm to 65nm processes with high speed and high on-chip density [3]. By comparison, low-end IoT edge devices are generally microcontroller-based, with low cost and more power-efficient SRAMs. The subsequent SoCs are frequently used in real-time operating systems and in wireless connectivity. The low-end IoT devices are currently designed using 65nm and 90nm mixed-signal processes and require ULP energy-efficient SRAM and embedded flash architectures [3].

1.9.3 Application of SRAM in Image-Processing Memory System

The object detection and tracking device basically use two types of architectures; one is the proximity sensor, and the other one is processing logic with the memory system.

The processing speed of tracking devices can be improved by a well-structured algorithm. However, to control the standby power and overall system power, a sub-threshold MOS architecture/hardware is required. Considering the fact that a tracking system requires a large amount of memory to store information, a high-capacity SRAM cell-based cache memory is essential. Due to its high accuracy and fast speed, the SRAM has been used in the image-processing and communications systems. The object detection and tracking system are shown in Figure 1.24 can be implemented in FPGAs using ULP SRAMs to store and track the change in movement by comparing present and previous frame information [89]. However, the systems designed for object detection and tracking consume a large amount of static power. The detection and tracking system requires a large amount of memory to store random moving object information, and the stored information is used to compare the successive frames to observe any change in motion. Therefore, a low-leakage and high-stability SRAM is required to minimize the power consumption and improve the accuracy of an object tracking system, respectively.

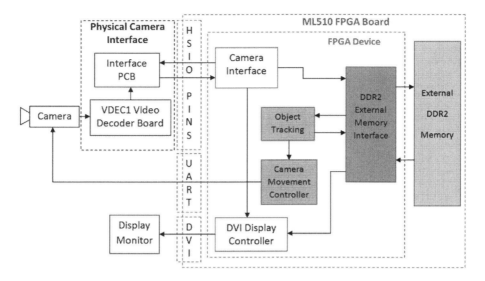

FIGURE 1.24 Object tracking system in FPGA. The object tracking uses high-speed DDR2 memory to store pixels in the form of bits and follow the movement of an object [88].

1.10 DESIGN CHALLENGES IN SRAM

Earlier, the work on SRAM focused on the performance and reliability of technology and density scaling. The use and implications of technology optimizations that are usually pursued for a broad range of high-volume and low-energy applications (e.g., mobile processors) have also begun to be investigated. However, the need to develop SRAM techniques to support energy-constrained applications such as biomedical devices, wireless sensor nodes, IoT edge devices, image processing, and mobile multimedia. Due to its heightening importance in digital systems, and its increased sensitivity to processing and manufacturing factors, SRAM design requires some level of coordination with technology development in order to be effective. As a result, low-power high-stability SRAM solutions must be compatible with industry methodologies, which are well suited for new technology development at the manufacturing level. For instance, optimal bitcell layout design depends on several manufacturing details. Accordingly, this work focuses on circuit techniques that are compatible with these approaches, particularly regarding the most advanced technologies. Chapter 3 contributes to identifying and solving some of the most critical issues faced by the subthreshold SRAMs.

1.10.1 Design Techniques

The book addresses some of the most critical issues of SRAM. For instance, to remove the constraints of read and write sizing conflict, this book suggests using separate paths for read and write. To make more stable and rugged SRAMs, this book describes some symmetric inverter pair SRAMs. To improve static noise margin, the following chapters also exhibit asymmetrical SRAM architectures. To reduce the power consumption, it demonstrates the subthreshold operation in addition to feedback cutting and power gating of SRAM cells. To improve the write speed, as well as write SNM, the book chapters also include a write assist technology. In this book, applications of ULP SRAMs in FPGAs, IoT edge devices, and image processing are discussed. These applications have different requirements, and hence, separate SRAM architectures are designed for each application. The FPGA block requires a high-performance, low-leakage SRAM; therefore, a feedback-controlled 10T SRAM is presented. The low-end IoT edge devices require high-stability energy-efficient SRAM; hence, a process-tolerant 10T SRAM is presented. For high-end IoT edge devices, a high-performance, high-stability, and ULP SRAM is needed; hence, a write-assist 10T SRAM is presented. However, for image-processing systems, a smaller size, a low-power, high-stability 8T SRAM is presented in the next chapters.

It is observed that the standby power and the cell stability are the key concern in the subthreshold SRAM designs to improve the reliability, yield, and susceptibility of portable electronic devices for application in medical devices, communication equipment, wearable portable devices, IoT edge devices, and image-processing systems. Cell stability is a foremost concern in the subthreshold region. The noise created from threshold variation, process variation, half-select issues, and multiple-bit errors reduces the stability of SRAM cells. Consequently, various cell architectures have been employed in literature to overcome these limitations for ULP applications [50,69,70,6,29]. Besides the advantages provided by these cells, there are some limitations like read stability, concerns of temperature variations in standby power, low WTP, and higher power delay product (PDP) in the subthreshold regime. To compensate for all these factors along with lower standby power and better cell stability, various ULP PVT-tolerant SRAMs are elucidated in this book. Following are the key points on SRAM architectures included in this book:

1. SRAMs operate in subthreshold regions and henceforth reduce the dynamic power
2. Feedback-cutting SRAMs are used for high SNMs
3. Power-gated SRAMs for low-power applications
4. Low read-write energy SRAMs for low-end IoT edge devices
5. High read and write speed SRAMs for high-end IoT edge devices
6. Low-energy, high-stability SRAM for object tracking systems

REFERENCES

1. Indicators-Historical Trends in Technical Themes. *Digital Systems, International Solid State Circuits Conference (ISSCC) Review Report 2015.* http://isscc.org/trends/, Accessed on November 22, 2021.
2. Solid State Technology. *Insight for Semiconductor Technology.* http://electroiq.com/, Accessed on November 22, 2021.

3. Designware IP for Internet of Things (IOT). https://www.synopsys.com/dw/doc.php/ds/o/internet_of_things_brochure.pdf
4. Shin-ichiro Abe, Yukinobu Watanabe, Nozomi Shibano, Nobuyuki Sano, Hiroshi Furuta, Masafumi Tsutsui, Taiki Uemura, and Takahiko Arakawa. Neutron-induced soft error analysis in mosfets from a 65nm to a 25 nm design rule using multi-scale monte carlo simulation method. In *Reliability Physics Symposium (IRPS), 2012 IEEE International*, pages SE–3. IEEE, 2012.
5. Peggy Abusaidi, Matt Klein, and Brian Philofsky. Virtex-5 FPGA system power design considerations. *Xilinx WP285 (v1. 0)*, February 14, 2008.
6. Sayeed Ahmad, Mohit Kumar Gupta, Naushad Alam, and Mohd Hasan. Single-ended schmitt-trigger-based robust low-power SRAM cell. *IEEE Transactions on Very Large Scale Integration (VLSI) Systems*, 24(8):2634–2642, 2016.
7. Muhammad A Alam. A critical examination of the mechanics of dynamic NBTI for pmosfets. In *Electron Devices Meeting, 2003. IEDM'03 Technical Digest. IEEE International*, pages 14.4.1–14.4.4. IEEE, 2003.
8. Robert Baumann. The impact of technology scaling on soft error rate performance and limits to the efficacy of error correction. In *Electron Devices Meeting, 2002. IEDM'02. International*, pages 329–332. IEEE, 2002.
9. Robert C Baumann. Radiation-induced soft errors in advanced semiconductor technologies. *IEEE Transactions on Device and Materials Reliability*, 5(3):305–316, 2005.
10. Ankur Beohar and Santosh K Vishvakarma. Performance enhancement of asymmetrical underlap 3d-cylindrical GAA-TFET with low spacer width. *Micro & Nano Letters*, 11(8):443–445, 2016.
11. Benton H Calhoun, Alice Wang, and Anantha Chandrakasan. Modeling and sizing for minimum energy operation in subthreshold circuits. *IEEE Journal of Solid-State Circuits*, 40(9):1778–1786, 2005.
12. Andrea Calimera, Mirko Loghi, Enrico Macii, and Massimo Poncino. Dynamic indexing: Leakage-aging co-optimization for caches. *IEEE Transactions on Computer-Aided Design of Integrated Circuits and Systems*, 33(2):251–264, 2014.
13. Ralph K Cavin, Paolo Lugli, and Victor V Zhirnov. Science and engineering beyond moore's law. *Proceedings of the IEEE*, 100(Special Centennial Issue):1720–1749, 2012.
14. Vikas Chandra and Robert Aitken. Impact of technology and voltage scaling on the soft error susceptibility in nanoscale CMOS. In *IEEE International Symposium on Defect and Fault Tolerance of VLSI Systems*, pages 114–122. IEEE, 2008.
15. Wen-Teng Chang, Cheng-Ting Shih, Jhao-Lin Wu, Shih-Wei Lin, Li-Gong Cin, and Wen-Kuan Yeh. Back-biasing to performance and reliability evaluation of UTBB FDSOI, bulk FinFETs, and SOI FinFETs. *IEEE Transactions on Nanotechnology*, 17(1):36–40, 2017.
16. David Choi, Kyu Choi, and John D Villasenor. New non-volatile memory structures for FPGA architectures. *IEEE Transactions on Very Large scale Integration (VLSI) Systems*, 16(7):874–881, 2008.
17. Altera Corporation. White paper: An analytical review of FPGA logic efficiency in stratix, virtex-ii & virtex-ii pro devices. *Technical report, Intel*, 2003.
18. Altera Corporation. White paper: FPGA architecture: Principles and progression. *Technical report*, 2021. https://www.eecg.utoronto.ca/~vaughn/papers/casm2021_arch_survey.pdf
19. Issam El Moukhtari, Vincent Pouget, Camille Larue, Frédéric Darracq, Dean Lewis, and Philippe Perdu. Impact of negative bias temperature instability on the single-event upset threshold of a 65nm SRAM cell. *Microelectronics Reliability*, 53(9):1325–1328, 2013.

20. Ivan S Esqueda. Confinement effects on radiation response of SOI FinFETs at the scaling limit. *IEEE Electron Device Letters*, 38(3):306–309, 2017.
21. Ming-Long Fan, Vita Pi-Ho Hu, Yin-Nien Chen, Pin Su, and Ching-Te Chuang. Variability analysis of sense amplifier for FinFET subthreshold SRAM applications. *IEEE Transactions on Circuits and Systems II: Express Briefs*, 59(12):878–882, 2012.
22. Michelle Fernandez and Peggy Abusaidi. Virtex-6 FPGA routing optimization design techniques. *White paper, Xilinx*, 2010. https://www.chipestimate.com/ip-docs/106/Xilinx-Virtex-6-FPGA-Routing-Optimization-Design-Techniques
23. Jacopo Franco, Ben Kaczer, Maria Toledano-Luque, Ph J Roussel, Jerome Mitard, LRagnarsson, Liesbeth Witters, Thomas Chiarella, Mitsuhiro Togo, Naoto Horiguchi, et al. Impact of single charged gate oxide defects on the performance and scaling of nanoscaled fets. In *Reliability Physics Symposium (IRPS), 2012 IEEE International*, pages 5A–4. IEEE, 2012.
24. Swaroop Ghosh and Kaushik Roy. Parameter variation tolerance and error resiliency: New design paradigm for the nanoscale era. *Proceedings of the IEEE*, 98(10):1718–1751, 2010.
25. Maxim S Gorbunov, Pavel S Dolotov, Andrey A Antonov, Gennady I Zebrev, Vladimir V Emeliyanov, Anna B Boruzdina, Andrey G Petrov, and Anastasia V Ulanova. Design of 65 nm CMOS SRAM for space applications: A comparative study. *IEEE Transactions on Nuclear Science*, 61(4):1575–1582, 2014.
26. R Habchi, C Salame, A Khoury, and P Mialhe. Temperature dependence of a silicon power device switching parameters. *Applied Physics Letters*, 88(15):153503, 2006.
27. Peter Hazucha, T Karnik, J Maiz, S Walstra, B Bloechel, J Tschanz, G Dermer, S Hareland, P Armstrong, and S Borkar. Neutron soft error rate measurements in a 90-nm CMOS process and scaling trends in SRAM from 0.25-/spl mu/m to 90-nm generation. In *Electron Devices Meeting, 2003. IEDM'03 Technical Digest. IEEE International*, pages 21–25. IEEE, 2003.
28. Ron Ho, Ken Mai, and Mark Horowitz. Efficient on-chip global interconnects. In *2003 Symposium on VLSI Circuits. Digest of Technical Papers (IEEE Cat. No. 03CH37408)*, pages 271–274. IEEE, 2003.
29. Aminul Islam and Mohd Hasan. Leakage characterization of 10t SRAM cell. *IEEE Transactions on Electron Devices*, 59(3):631–638, 2012.
30. Shailendra Jain, Surhud Khare, Satish Yada, V Ambili, Praveen Salihundam, Shiva Ramani, Sriram Muthukumar, M Srinivasan, Arun Kumar, Shasi Kumar Gb, et al. A 280mv-to-1.2 v wide-operating-range ia-32 processor in 32nm CMOS. In *2012 IEEE International Solid-state Circuits Conference*, pages 66–68. IEEE, 2012.
31. Kjell O Jeppson and Christer M Svensson. Negative bias stress of MOS devices at high electric fields and degradation of MNOS devices. *Journal of Applied Physics*, 48(5):2004–2014, 1977.
32. Hai Jiang, Xiaoyan Liu, Nuo Xu, Yandong He, Gang Du, and Xing Zhang. Investigation of self-heating effect on hot carrier degradation in multiple-fin SOI FinFETs. *IEEE Electron Device Letters*, 36(12):1258–1260, 2015.
33. GF Jiao, ZX Chen, HY Yu, XY Huang, DM Huang, N Singh, GQ Lo, D-L Kwong, and Ming-Fu Li. Experimental studies of reliability issues in tunneling field-effect transistors. *IEEE Electron Device Letters*, 31(5):396–398, 2010.
34. Ben Kaczer, Tibor Grasser, Ph J Roussel, Jacopo Franco, Robin Degraeve, L-A Ragnarsson, Eddy Simoen, Guido Groeseneken, and Hans Reisinger. Origin of NBTI variability in deeply scaled PFETS. In *Reliability Physics Symposium (IRPS), 2010 IEEE International*, pages 26–32. IEEE, 2010.

35. Ben Kaczer, S Mahato, V Valduga de Almeida Camargo, M Toledano-Luque, Ph J Roussel, T Grasser, Francky Catthoor, P Dobrovolny, P Zuber, G Wirth, et al. Atomistic approach to variability of bias-temperature instability in circuit simulations. In *Reliability Physics Symposium (IRPS), 2011 IEEE International*, pages XT–3. IEEE, 2011.
36. Sung-Mo Kang and Yusuf Leblebici. *CMOS Digital Integrated Circuits*. Tata McGrawHill Education, 2003.
37. Amy V Kauppila, Bharat L Bhuva, TD Loveless, S Jagannathan, NJ Gaspard, JS Kauppila, Lloyd W Massengill, SJ Wen, R Wong, GL Vaughn, et al. Effect of negative BIAS temperature instability on the single event upset response of 40 nm flip-flops. *IEEE Transactions on Nuclear Science*, 59(6):2651–2657, 2012.
38. John P Keane. *On-chip circuits for characterizing transistor aging mechanisms in advanced CMOS technologies*, 2010.
39. Navid Khoshavi, Hamid R Zarandi, and Mohammad Maghsoudloo. Control-flow error detection using combining basic and program-level checking in commodity multi-core architectures. In *Industrial Embedded Systems (SIES), 2011 6th IEEE International Symposium on*, pages 103–106. IEEE, 2011.
40. Navid Khoshavi, Hamid R Zarandi, and Mohammad Maghsoudloo. Control-flow error recovery using commodity multi-core architecture features. In *On-Line Testing Symposium (IOLTS), 2011 IEEE 17th International*, pages 190–191. IEEE, 2011.
41. Jin Sang Kim, Ik Joon Chang, et al. We-quatro: Radiation-hardened SRAM cell with parametric process variation tolerance. *IEEE Transactions on Nuclear Science*, 64(9):2489–2496, 2017.
42. Kidong Kim, Ohseob Kwon, Jihyun Seo, and Taeyoung Won. Nanoscale device modeling and simulation: Fin field-effect transistor (FinFET). *Japanese Journal of Applied Physics*, 43(6S):3784, 2004.
43. Kinam Kim and Gwan-Hyeob Koh. Future memory technology including emerging new memories. In *2004 24th International Conference on Microelectronics (IEEE Cat. No. 04TH8716)*, volume 1, pages 377–384. IEEE, 2004.
44. Nam Sung Kim, Todd Austin, David Baauw, Trevor Mudge, Krisztian Flautner, Jie S' Hu, Mary Jane Irwin, Mahmut Kandemir, and Vijaykrishnan Narayanan. Leakage current: Moore's law meets static power. *Computer*, 36(12):68–75, 2003.
45. Tony Tae-Hyoung Kim and Zhi Hui Kong. Impact analysis of NBTI/PBTI on SRAM v MIN and design techniques for improved SRAM v MIN. *JSTS: Journal of Semiconductor Technology and Science*, 13(2):87–97, 2013.
46. Alicia Klinefelter, Nathan E Roberts, Yousef Shakhsheer, Patricia Gonzalez, Aatmesh Shrivastava, Abhishek Roy, Kyle Craig, Muhammad Faisal, James Boley, Seunghyun Oh, et al. 21.3 a 6.45 μw self-powered IOT SOC with integrated energy-harvesting power management and ULP asymmetric radios. In *2015 IEEE International Solid-State Circuits Conference-(ISSCC) Digest of Technical Papers*, pages 1–3. IEEE, 2015.
47. Meishoku Koh, Wataru Mizubayashi, Kunihiko Iwamoto, Hideki Murakami, Tsuyoshi Ono, Morikazu Tsuno, Tatsuyoshi Mihara, Kentaro Shibahara, Seiichi Miyazaki, and Masataka Hirose. Limit of gate oxide thickness scaling in mosfets due to apparent threshold voltage fluctuation induced by tunnel leakage current. *IEEE Transactions on Electron Devices*, 48(2):259–264, 2001.
48. Oleg Kononchuk and B-Y Nguyen. *Silicon-on-Insulator (SOI) Technology: Manufacture and Applications*. Elsevier, 2014.
49. Selahaddin Halil Kükner. *Bias Temperature Instability in CMOS Digital Circuits from Planar to FinFET Nodes*, 2015.
50. Jaydeep P Kulkarni and Kaushik Roy. Ultralow-voltage process-variation-tolerant schmitt-trigger-based SRAM design. *IEEE Transactions on Very Large Scale Integration (VLSI) Systems*, 20(2):319–332, 2011.

Introduction

51. Volkan Kursun and Eby G Friedman. *Multi-voltage CMOS Circuit Design*. 2006. https://www.wiley.com/en-in/Multi+voltage+CMOS+Circuit+Design-p-9780470010235
52. CB Kushwah and Santosh Kumar Vishvakarma. A single-ended with dynamic feedback control 8t subthreshold SRAM cell. *IEEE Transactions on Very Large Scale Integration (VLSI) Systems*, 24(1):373–377, 2015.
53. Jian Li and Alper Buyuktosunoglu. Session details: Memory system design. In *Proceedings of the 16th ACM/IEEE International Symposium on Low Power Electronics and Design*, ISLPED '10. Association for Computing Machinery, 2010.
54. Shaoshan Liu, Liangkai Liu, Jie Tang, Bo Yu, Yifan Wang, and Weisong Shi. Edge computing for autonomous driving: Opportunities and challenges. *Proceedings of the IEEE*, 107(8):1697–1716, 2019.
55. S Mahapatra, V Huard, A Kerber, V Reddy, S Kalpat, and A Haggag. Universality of nbti-from devices to circuits and products. In *Reliability Physics Symposium, 2014 IEEE International*, pages 3B–1. IEEE, 2014.
56. Souvik Mahapatra, Muhammad A Alam, P Bharath Kumar, TR Dalei, Dhanoop Varghese, and Dipankar Saha. Negative bias temperature instability in CMOS devices. *Microelectronic Engineering*, 80:114–121, 2005.
57. Elie Maricau and Georges Gielen. *Analog IC Reliability in Nanometer CMOS*. Springer, 2013.
58. Erik Jan Marinissen, Betty Prince, D Keltel-Schulz, and Yervant Zorian. Challenges in embedded memory design and test. In *Design, Automation and Test in Europe*, pages 722–727. IEEE, 2005.
59. N Mehta. Xilinx redefines power, performance, and design productivity with three innovative 28 nm FPGA families: Virtex-7, kintex-7, and artix-7 devices.—*xilinx inc.—white paper wp373 (v1. 4) October 15, 2012*, p. 10, 2012. https://www.chipestimate.com/Xilinx/Xilinx-Redefines-Power-Performance-and-Design-Productivity-with-Three-Innovative-28-nm-FPGA-Families-Virtex-7-Kintex-7-and-Artix-7-Devices-/White-Paper/95
60. Nick Mehta. Xilinx 7 series fpgas: The logical advantage. *Xilinx WP405*, 2012. https://www.techonline.com/tech-papers/xilinx-7-series-fpgas-the-logical-advantage/
61. Tomohisa Mizuno, J Okumtura, and Akira Toriumi. Experimental study of threshold voltage fluctuation due to statistical variation of channel dopant number in mosfet's. *IEEE Transactions on Electron Devices*, 41(11):2216–2221, 1994.
62. Baker S Mohammad, Hani Saleh, and Mohammed Ismail. Design methodologies for yield enhancement and power efficiency in SRAM-based socs. *IEEE Transactions on Very Large Scale Integration (VLSI) Systems*, 23(10):2054–2064, 2014.
63. Dheeraj Mohata, Bijesh Rajamohanan, Theresa Mayer, Mantu Hudait, Joel Fastenau, Dmitri Lubyshev, Amy WK Liu, and Suman Datta. Barrier-engineered arsenide—antimonide heterojunction tunnel fets with enhanced drive current. *IEEE Electron Device Letters*, 33(11):1568–1570, 2012.
64. Gordon E Moore. Cramming more components onto integrated circuits. *Proceedings of the IEEE*, 86(1):82–85, 1998.
65. Saibal Mukhopadhyay, Hamid Mahmoodi, and Kaushik Roy. Modeling of failure probability and statistical design of SRAM array for yield enhancement in nanoscaled CMOS. *IEEE Transactions on Computer-Aided Design of Integrated Circuits and Systems*, 24(12):1859–1880, 2005.
66. Ayhan A Mutlu and Mahmud Rahman. Statistical methods for the estimation of process variation effects on circuit operation. *IEEE Transactions on Electronics Packaging Manufacturing*, 28(4):364–375, 2005.
67. Mehrdad Nourani and Arun Radhakrishnan. Testing on-die process variation in nanometer VLSI. *IEEE Design & Test of Computers*, 23(6), 2006.

68. Y Ohnari, AA Khan, A Dutta, M Miura-Mattausch, and HJ Mattausch. Die-to-die and within-die variation extraction for circuit simulation with surface-potential compact model. In *Microelectronic Test Structures (ICMTS), 2013 IEEE International Conference on*, pages 146–150. IEEE, 2013.
69. Soumitra Pal and Aminul Islam. Variation tolerant differential 8t SRAM cell for ultralow power applications. *IEEE Transactions on Computer-Aided Design of Integrated Circuits and Systems*, 35(4):549–558, 2015.
70. Soumitra Pal and Aminul Islam. 9-t SRAM cell for reliable ultralow-power applications and solving multibit soft-error issue. *IEEE Transactions on Device and Materials Reliability*, 16(2):172–182, 2016.
71. Georgios D Panagopoulos and Kaushik Roy. A three-dimensional physical model for Vth variations considering the combined effect of NBTI and RDF. *IEEE Transactions on Electron Devices*, 58(8):2337–2346, 2011.
72. Sang Phill Park, Kunhyuk Kang, and Kaushik Roy. Reliability implications of biastemperature instability in digital ICS. *IEEE Design & Test of Computers*, 26(6), 2009.
73. Harsh N Patel, Farah B Yahya, and Benton H Calhoun. Optimizing SRAM bitcell reliability and energy for IOT applications. In *2016 17th International Symposium on Quality Electronic Design (ISQED)*, pages 12–17. IEEE, 2016.
74. Gordon Patrick and A Gattani. Memory plays a vital role in building the connected world. *Electronic Design: Hong Kong, China*, 2015.
75. Somnath Paul, Saibal Mukhopadhyay, and Swarup Bhunia. A variation-aware preferential design approach for memory-based reconfigurable computing. *IEEE Transactions on Very Large Scale Integration (VLSI) Systems*, 22(12):2449–2461, 2014.
76. Yogesh K Ramadass and Anantha P Chandrakasan. Minimum energy tracking loop with embedded DC-DC converter delivering voltages down to 250mv in 65nm CMOS. In *2007 IEEE International Solid-State Circuits Conference. Digest of Technical Papers*, pages 64–587. IEEE, 2007.
77. Daniele Rossi, Martin Omaña, Cecilia Metra, and Alessandro Paccagnella. Impact of bias temperature instability on soft error susceptibility. *IEEE Transactions on Very Large Scale Integration (VLSI) Systems*, 23(4):743–751, 2015.
78. Kaushik Roy and Sharat C Prasad. *Low-power CMOS VLSI Circuit Design*. John Wiley & Sons, 2009.
79. Takayasu Sakurai and A Richard Newton. Alpha-power law mosfet model and its applications to CMOS inverter delay and other formulas. *IEEE Journal of Solid-State Circuits*, 25(2):584–594, 1990.
80. Sachin S Sapatnekar. Overcoming variations in nanometer-scale technologies. *IEEE Journal on Emerging and Selected Topics in Circuits and Systems*, 1(1):5–18, 2011.
81. Dieter K Schroder. Negative bias temperature instability: What do we understand? *Microelectronics Reliability*, 47(6):841–852, 2007.
82. Dieter K Schroder and Jeff A Babcock. Negative bias temperature instability: Road to cross in deep submicron silicon semiconductor manufacturing. *Journal of Applied Physics*, 94(1):1–18, 2003.
83. Brian D Sierawski, Kevin M Warren, Robert A Reed, Robert A Weller, Marcus M Mendenhall, Ronald D Schrimpf, Robert C Baumann, and Vivian Zhu. Contribution of low-energy (¡¡ 10 mev) neutrons to upset rate in a 65 nm SRAM. In *Reliability Physics Symposium (IRPS), 2010 IEEE International*, pages 395–399. IEEE, 2010.
84. Mahmut E Sinangil and Anantha P Chandrakasan. Application-specific SRAM design using output prediction to reduce bit-line switching activity and statistically gated sense amplifiers for up to 1.9*times* lower energy/access. *IEEE Journal of Solid-State Circuits*, 49(1):107–117, 2013.

85. Ashish K Singh, Ku He, Constantine Caramanis, and Michael Orshansky. Modeling and optimization techniques for yield-aware SRAM post-silicon tuning. *IEEE Transactions on Computer-Aided Design of Integrated Circuits and Systems*, 33(8):1159–1167, 2014.
86. Pooran Singh and Santosh Kumar Vishvakarma. Device/circuit/architectural techniques for ultra-low power FPGA design. *Microelectronics and Solid State Electronics*, 2(A):1–15, 2013.
87. Prashant Singh, Eric Karl, David Blaauw, and Dennis Sylvester. Compact degradation sensors for monitoring nbti and oxide degradation. *IEEE Transactions on Very Large Scale Integration (VLSI) Systems*, 20(9):1645–1655, 2012.
88. Sanjay Singh, Chandra Shekhar, and Anil Vohra. Real-time FPGA-based object tracker with automatic pan-tilt features for smart video surveillance systems. *Journal of Imaging*, 3(2), 2017.
89. Sanjay Singh, Chandra Shekhar, and Anil Vohra. Real-time FPGA-based object tracker with automatic pan-tilt features for smart video surveillance systems. *Journal of Imaging*, 3(2):18, 2017.
90. James H Stathis, Souvik Mahapatra, and Tibor Grasser. Controversial issues in negative bias temperature instability. *Microelectronics Reliability*, 81:244–251, 2018.
91. Akhil Sudarsanan, Sankatali Venkateswarlu, and Kaushik Nayak. Impact of fin line edge roughness and metal gate granularity on variability of 10-nm node SOI n-FinFET. *IEEE Transactions on Electron Devices*, 66(11):4646–4652, 2019.
92. K Takeuchi, T Fukai, T Tsunomura, AT Putra, A Nishida, S Kamohara, and T Hiramoto. Understanding random threshold voltage fluctuation by comparing multiple fabs and technologies. In *2007 IEEE International Electron Devices Meeting*, pages 467–470. IEEE, 2007.
93. Taiki Uemura, Takashi Kato, Hideya Matsuyama, and Masanori Hashimoto. Soft-error in SRAM at ultra-low voltage and impact of secondary proton in terrestrial environment. *IEEE Transactions on Nuclear Science*, 60(6):4232–4237, 2013.
94. Osman S Unsal, James W Tschanz, Keith Bowman, Vivek De, Xavier Vera, Antonio Gonzalez, and Oguz Ergin. Impact of parameter variations on circuits and microarchitecture. *Ieee Micro*, 26(6):30–39, 2006.
95. Rangharajan Venkatesan, Vivek J Kozhikkottu, Mrigank Sharad, Charles Augustine, Arijit Raychowdhury, Kaushik Roy, and Anand Raghunathan. Cache design with domain wall memory. *IEEE Transactions on Computers*, 65(4):1010–1024, 2015.
96. P-F Wang, K Hilsenbeck, Th Nirschl, M Oswald, Ch Stepper, M Weis, D SchmittLandsiedel, and W Hansch. Complementary tunneling transistor for low power application. *Solid-State Electronics*, 48(12):2281–2286, 2004.
97. Pei-Yu Wang and Bing-Yue Tsui. Experimental demonstration of p-channel germanium epitaxial tunnel layer (etl) tunnel fet with high tunneling current and high on/off ratio. *IEEE Electron Device Letters*, 36(12):1264–1266, 2015.
98. Raymond Wang, James Dunkley, Thomas A DeMassa, and Lawrence F Jelsma. Threshold voltage variations with temperature in MOS transistors. *IEEE Transactions on Electron Devices*, 18(6):386–388, 1971.
99. Xingsheng Wang, Andrew R Brown, Binjie Cheng, and Asen Asenov. Statistical variability and reliability in nanoscale FinFETs. In *Electron Devices Meeting (IEDM), 2011 IEEE InternationaDl*, pages 5–4. IEEE, 2011.
100. Yao Wang, Marius Enachescu, Sorin Dan Cotofana, and Liang Fang. Variation tolerant on-chip degradation sensors for dynamic reliability management systems. *Microelectronics Reliability*, 52(9):1787–1791, 2012.
101. Linda Wilson. International technology roadmap for semiconductors (itrs). *Semiconductor Industry Association*, 1, 2013.

102. Martin Wirnshofer. *Variation-Aware Adaptive Voltage Scaling for Digital CMOS Circuits*. Springer series in Advanced Microelectronics. Springer, 2013.
103. Xin Wu, Prabhuram Gopalan, and Greg Lara. Xilinx next generation 28 nm FPGA technology overview. *by Xilinx white paper*, 2010. https://docs.xilinx.com/v/u/en-US/wp312_Next_Gen_28_nm_Overview
104. Zhao Yuanfu, Yue Suge, Zhao Xinyuan, Lu Shijin, Bian Qiang, Wang Liang, and Sun Yongshu. Single event soft error in advanced integrated circuit. *Journal of Semiconductors*, 36(11):111001, 2015.
105. Kevin Zhang. F1: Embedded memory design for nano-scale VLSI systems. In *2008 IEEE International Solid-State Circuits Conference-Digest of Technical Papers*, pages 650–651. IEEE, 2008.
106. Wei Zhang, Ki Chul Chun, and Chris H Kim. A write-back-free 2t1d embedded dram with local voltage sensing and a dual-row-access low power mode. *IEEE Transactions on Circuits and Systems I: Regular Papers*, 60(8):2030–2038, 2013.
107. Xi Zhang, Daniel Connelly, Hideki Takeuchi, Marek Hytha, Robert J Mears, and TsuJae King Liu. Comparison of SOI versus bulk FinFET technologies for 6t-SRAM voltage scaling at the 7-/8-nm node. *IEEE Transactions on Electron Devices*, 64(1):329–332, 2016.
108. Yanqing Zhang, Fan Zhang, Yousef Shakhsheer, Jason D Silver, Alicia Klinefelter, Manohar Nagaraju, James Boley, Jagdish Pandey, Aatmesh Shrivastava, Eric J Carlson, et al. A batteryless 19 *mu* w mics/ism-band energy harvesting body sensor node soc for EXG applications. *IEEE Journal of Solid-State Circuits*, 48(1):199–213, 2012.
109. Fengxia Zhao, Linna Zhang, Qinghai Wang, and Zhuangde Jiang. Impact of line edge roughness and linewidth roughness on critical dimension variation. In *Computer Science and Automation Engineering (CSAE), 2012 IEEE International Conference on*, volume 3, pages 475–479. IEEE, 2012.

2 Design Metrics for Embedded SRAM

Memory has been the driving force behind the rapid growth of complementary metal-oxide semiconductor (CMOS) technology for the past few decades. It all started with the first 1-kb dynamic random access memory (DRAM) developed by Intel in the 1970s to the current generation in which DRAM capacities have reached beyond 1 Gb [1]. The arrival of virtual memory in personal computers (PCs) contributed to the hierarchical structure of various kinds of memory, ranging from small-capacity, fast, but more costly cache memories to large-capacity, slower, but more affordable magnetic and optical storage. The pyramid hierarchy of memory types in a PC is shown in Figure 2.1, reflecting the growing speed and cost per bit as we move from the bottom, Level 5 (L5) remote secondary storage, to the topmost register storage, Level (L0). The introduction of the memory hierarchy is a fundamental consequence of maintaining the random-access memory abstraction and practical limits on the cost and the power consumption.

The mounting gap between the microprocessor cycle time and DRAM access time imposed the introduction of several levels of caching in modern data processors. In

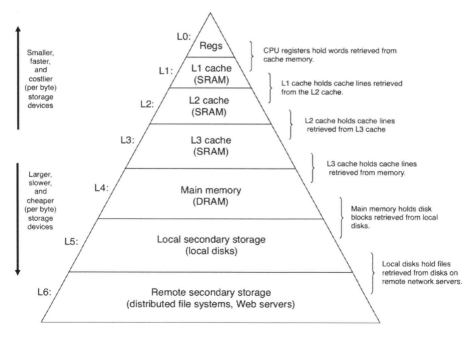

FIGURE 2.1 Memory hierarchy of a personal computer [1].

DOI: 10.1201/9781003213451-2

PCs, such levels are often represented by L1 and L2 on-chip embedded SRAM cache memories. As the speed gap between the microprocessor, memory, and mass storage continues to widen, deeper memory hierarchies have been introduced in high-end server microprocessors. Furthermore, to improve the requirement of access time in high-end microprocessor cache, Level 3 (L3) SRAM has been introduced [2]. This can help PCs meet the time gap between the processor and memory. However, this may increase the overall cost of a PC, but due to increased demand for high-end processors, the introduction of an L3 cache is required.

2.1 CONVENTIONAL SRAM ARCHITECTURE: 6T AND RD8T SRAM BITCELLS

L1, L2, and L3 in the memory hierarchy shown in Figure 2.1 are cache memories that are implemented using conventional SRAM bitcells like 6-transistor (6T) and read-decoupled 8T (RD8T). The CMOS-based circuit diagram of 6T and RD8T SRAM bitcells are shown in Figures 2.2a and 2.2b, respectively. Each bitcell is capable of storing a single bit of information. It provides nondestructive read operation, write capability, and data storage as long as the SRAM bitcell is powered up. A typical SRAM memory architecture along with a simple schematic illustration of column circuitry is shown in Figure 1.2. The data storage element consists of 6T or 8T SRAM bitcells organized in an array of rows and columns. Here, each bitcell is capable of storing one bit of binary data. Also, each 6T bitcell shares a shared connection with other cells in the same row called wordline (WL) and an additional common signal with the other cells in the same column called column select (CS). In this structure, there are N rows and M columns. Thus, the total number of memory cells in this array is M × N. The memory array of size M × N is designed using a 6T SRAM bitcell with differential read sensing through a sense amplifier (SA) placed at each column. In addition, to improve the readability and read margins, an RD8T SRAM array is used most of the time. The 6T SRAM cell shown in Figure 2.2a contains a pair of cross-coupled inverters (M1–M4), which hold the data, and a pair of access transistors (M5–M6) to initiate the read or write operation. However, an RD8T shown in Figure 2.2b is designed to isolate the read-write

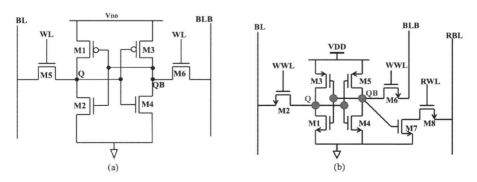

FIGURE 2.2 (a) Conventional 6T SRAM. (b) Standard RD8T SRAM.

Design Metrics for Embedded SRAM

path and to improve the read noise margin (RNM) in the read operation. To access a particular memory cell, the corresponding CS and WL must be activated. The row and column decoders are employed to accomplish row and column selection operations, respectively. The data are written into the SRAM bitcell by applying the logic bit value and its complemented value onto the bit lines and then triggering the wordline (WL for 6T and write wordline (WWL) for 8T). Additionally, the column circuitry also consists of the global read/write circuitry, the bitline (*BL*) sensing, and the column multiplexers as shown in Figure 1.2.

An SRAM cell can store one bit of data. It comprises two back-to-back connected inverters that form a latch and two access transistors. In 6T, a WL select signal is connected with the access transistors to allow reading or writing to the cell by connecting bitlines to the cell storage nodes. An SRAM offers the following basic properties:

- **Retention/Hold:** Data can be indefinitely stored in an SRAM cell as long as it is powered.
- **Read:** An SRAM cell can convey the data effectively without destroying the data itself.
- **Write:** The desired bit can be written on the cell regardless of what data are present in it.

2.1.1 6T SRAM READ OPERATION

During the read operation in the 6T SRAM cell, both the bitline (*BL*) and bitline bar (*BLB*) are pre-charged to V_{DD} as shown in Figure 2.3. The precharge signal (PRE) is asserted to '0,' which makes both bitlines be set to the V_{DD}. For the read operation, *WL* is asserted to '1,' and PRE switches from logic 0 to logic 1. Once the BLs are removed from power line, one of the BLs will tend to discharge due to the potential difference in the *BL* and the storage node with a logic 0 value. The difference between the BLs (Δ_{BIT}) is read by the differential sense amplifier (SA) placed at the bottom of the array.

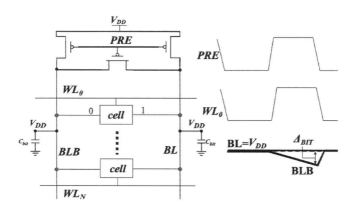

FIGURE 2.3 Read operation setup for a 6T SRAM.

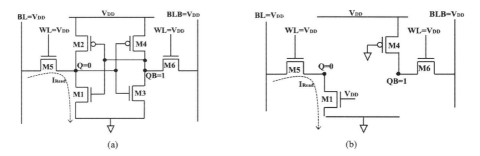

FIGURE 2.4 (a) Read '0' operation setup for 6T SRAM. (b) Simplified circuit for read '0' operation.

A read operation in a 6T SRAM bitcell is performed by pulling the WL high, giving the BLs access to the storage nodes (Q and QB) and the latch. The storage nodes are complementary to each other. Prior to the read operation, the BLs are precharged to V_{DD} equally, during a read '0' operation, as shown in Figures 2.4a and 2.4b, storage node Q holds the value '0', whereas its complementary node holds '1'. When the access transistors are turned ON by pulling the WL HIGH, bitline *BLB* stays high as storage node QB stays high during the read '0' operation, and bitline *BL* is pulled down due to the discharges through access transistor M5 and pull-down device 'M1'. The voltage difference between the BLs is sensed by the differential SA and read at the output driver. Similarly, during a 'read 1' operation, Q holds '1' and QB holds '0'; when the WL is pulled high, the bitline *BLB* is discharged through access transistor M6 and pull-down transistor 'M3,' and *BL* remains the same due to the pull-up logic, M6 goes to the saturation region while M3 operates in the triode region. The change in voltage in *BL* and *BLB* is detected as stored logic '1' in the bitcell by the SAs.

The read current and read access time are evaluated by the model shown in Figure 2.5 and Equation 2.1, where T_{access} is the read access time and Δ_{BIT} is the bit differential voltage. The read and write access times depend on the parasitic and junction capacitance values of the access transistors. There are three capacitances associated with metal-oxide semiconductor (MOS) devices, namely, C_j, C_{ov}, and C_m,

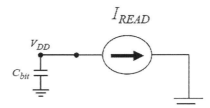

FIGURE 2.5 Model for read current and read acess times.

Design Metrics for Embedded SRAM

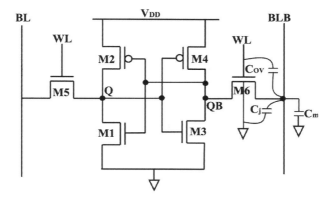

FIGURE 2.6 The capacitances on the *BL* and *BLB* of an SRAM array.

mentioned in Equation 2.2 and shown in Figure 2.6, where N_{row} is the total number of rows in the SRAM.

$$C_{bit}\, dV_{bl} + I_{READ} = 0$$

$$C_{bit} \frac{\Delta_{BIT}}{I_{READ}} = T_{access} \quad (2.1)$$

$$T_{access} = \text{Access Time}$$

$$\Delta_{BIT} = bit - differential$$

$$C_{bit} = N_{row}(C_j + C_{ov} + C_m)$$

$$C_j = \text{junction cap of each access device} \quad (2.2)$$

$$C_{ov} = \text{gate to drain overlap cap of each access device}$$

$$C_m = \text{mental cap of per cell}$$

For evaluating the read voltage shown in Figure 2.7, V_{READ}, the voltage at the storage node, which is at '0,' it is observed that M5 is in saturation mode, and the read current (I_{READ}) is determined by Equation 2.3. Using Equations 2.1, 2.2, and 2.3, the access time is derived as stated in Equation 2.4.

$$I_{READ} = \mu_n C_{ox} \frac{W_{M5}}{2L_{M5}}(V_{DD} - V_{READ} - V_{th})^2 = 0.5\beta_{M5}(V_{DD} - V_{READ} - V_{th})^2 \quad (2.3)$$

$$T_{ACCESS} = C_{bit} \frac{\Delta_{BIT}}{0.5\beta_{M5}(V_{DD} - V_{READ} - V_{th})^2} \quad (2.4)$$

Similarly, the read voltage is determined by the current-voltage equations through M5 (in saturation) and M1 (in linear) as shown in Figure 2.8 and Equation 2.5. The

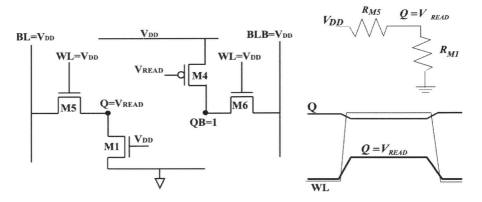

FIGURE 2.7 Evaluation of the read voltage of a 6T SRAM.

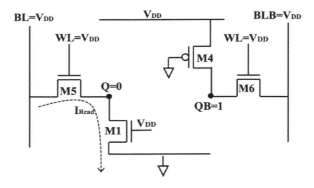

FIGURE 2.8 Model for read voltage estimation.

β ratio of M1 and M5 is the deciding factor to determine the read voltage across the storage node. The higher the $\beta_{pd\text{-}ax}$ (pull-down-to-access) value, the lower the read voltage and vice versa. To withstand a nondestructive read operation in a 6T SRAM, the read voltage V_{READ} should be less than the trip voltage, V_{trip}, as shown in Figure 2.9.

$$0.5\beta_{M5}(V_{DD} - V_{READ} - V_{th})^2 = \beta_{M1}[(V_{DD} - V_{th}) - 0.5V_{READ}]V_{READ}$$

$$(1 + \beta_{M1}/\beta_{M5})V_{READ}^2 - 2(1 + \beta_{M1}/\beta_{M5})(V_{DD} - V_{th})V_{READ} + (V_{DD} - V_{th})^2 = 0$$

$$V_{READ} = (V_{DD} - V_{th})\left(1 - \sqrt{1 - 1/(1 + \beta_{pd-ax})}\right) \tag{2.5}$$

Furthermore, the read margin can be evaluated as a difference between V_{trip} (trip voltage) and V_{READ} (read voltage), where the trip voltage is defined as the switching threshold voltage (Equation 2.6) of the CMOS inverter.

$$Read\,Margin = V_{trip} - V_{READ}$$

Design Metrics for Embedded SRAM

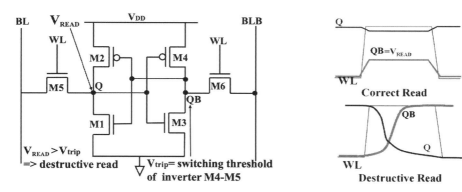

FIGURE 2.9 Correct and destructive read operation in 6T SRAM.

$$V_{trip} = \frac{V_{DD} - |V_{thp}| + \sqrt{\beta_{M3}/\beta_{M4}}\, V_{thn}}{1 + \sqrt{\beta_{M3}/\beta_{M4}}} \quad \begin{array}{l} \text{stronger pull-down} \Rightarrow \text{lower } V_{trip} \\ \text{stronger pull-up} \Rightarrow \text{higher } V_{trip} \end{array}$$

$$= \frac{V_{DD} - |V_{thp}| + \sqrt{\beta_{ratio-pd-pup}}\, V_{thn}}{1 + \sqrt{\beta_{ratio-pd-pup}}} \qquad (2.6)$$

$$V_{READ} = (V_{DD} - V_{thn})\left(1 - \sqrt{1 - \frac{1}{1 + \beta_{ratio-pd-ax}}}\right)$$

The read margin value is improved by reducing the read voltage. A stronger pull-down device reduces V_{READ}; however, a stronger access device increases V_{READ} as shown in Figure 2.10a. The trip voltage versus the pull-down–pull-up beta ratio is shown in Figure 2.10b. Figure 2.10b shows that a stronger pull-up device increases V_{trip} and that a weaker pull-down device increases V_{trip}.

Similarly, in an RD8T SRAM, prior to the read operation, *RBL*, or read bitline, is pulled high or precharged to V_{DD}, and *WWL* is pulled low for the read operation, which separates the read logic to write logic. The read operation is performed by enabling the read wordline (*RWL*) signal. As shown in Figure 2.11, the dedicated read port temporarily decouples the read path from the storage nodes during the read operation, therefore enabling a nondestructive read operation since *WWL* is pulled low. During a 'read 0' operation, Q holds logic 0, and QB holds logic 1, which is maintained by the pull-up logic (M5). This, in turn, turns on the pull-down logic and drains the *RBL* voltage to the ground through M7 and M8, which is detected by the inverter-based SAs as logic 0. During a read '1' operation, Q holds '1,' and 'QB' is pulled down to '0' by pull-down logic (M4), and M7 stays 'off,' thereby having no change in voltage in *RBL* and stays at precharged voltage.

Here, M7 and M8 both are in saturation mode, and the read voltage will be across the M7–M8, which will not affect the trip point of the M1–M3 CMOS inverter. Thus,

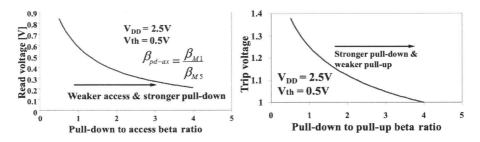

FIGURE 2.10 (a) Read voltage to pull-down–access beta ratio. (b) Trip voltage versus pull-down–pull-up beta ratio.

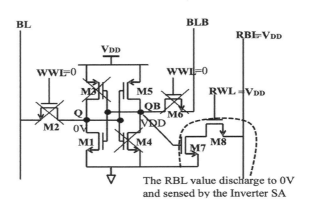

FIGURE 2.11 Read '1' operation setup for an RD8T SRAM bitcell.

the read margin of an SRAM would be equal to the trip point or trip voltage of the opposite CMOS inverter.

2.1.2 Write Operation in SRAM

In a 6T SRAM, the write operation is performed by controlling the bitline voltages and forcing them into storage nodes Q and QB. A Boolean value is written (or stored) into the 6T SRAM bitcell by initializing the bitlines to the required voltage and activating the access transistors. As shown in Figure 2.12, during a write '0' operation, to write a logic 0 and its complementary logic 1 into the bitcell, we pull down *BL* to VSS (logic 0) and pull its complimentary *BLB* to V_{DD} (logic 1). When the *WL* is activated voltage at Q is discharged through the access transistor M5, and the write operation is completed when storage node Q is pulled down to low or below the trip voltage of inverter M3–M4. Similarly, a write '1' is performed by bringing down *BLB* to 0, pulling up *BL* to V_{DD}, and discharging voltage at QB and flipping the data. In addition, as shown in Figure 2.13, during the write operation, the pull-up transistor supplies electrical charge to the storage node Q; therefore, if the charging

Design Metrics for Embedded SRAM 53

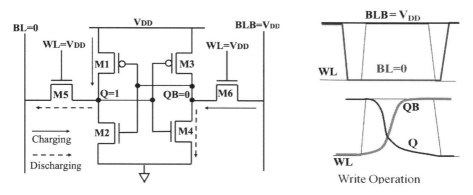

FIGURE 2.12 Write '0' operation setup for a 6T SRAM cell.

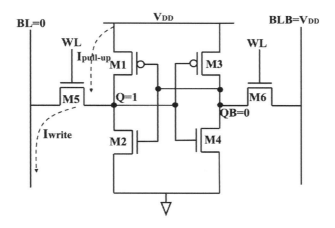

FIGURE 2.13 Write failure due to higher pull-up current ($I_{pull-up}$) as compared to discharging write current (I_{write}).

current $I_{Pull-up}$ is greater than the discharging current of the access transistor, voltages at storage nodes are not flipped and a write failure occurs.

The write operation for 6T and 8T SRAM is similar, initially both the bitlines are precharged to V_{DD} and for writing information one of *BL* and *BLB* would be discharged to '0' as shown in Figure 2.14a. The major key points for a successful write operation in both 6T and RD8T SRAM are the following:

- The major writing process is discharging the node storing '1' to '0.'
- During this discharging process, the ON pull-up p-type MOSFET (PMOS) fights with the ON access device.
- Need to make sure the node storing '1' can be reduced below the switching threshold of the other inverter.

$$0.5\beta_{M4}(V_{DD}-V_{th})^2 = \beta_{M6}[(V_{DD}-V_{th})-0.5V_Q]V_Q$$

$$V_Q^2 - 2(V_{DD}-V_{th})V_Q + \beta_{M4}/\beta_{M6}(V_{DD}-V_{th})^2 = 0$$

$$V_Q = (V_{DD}-V_{th})\left(1-\sqrt{1-\beta_{ratio-pup-ax}}\right) \qquad (2.7)$$

To determine the write voltage V_Q as shown in Equation 2.7, we have estimated that M4 is operating in saturation and M6 in the linear region. The goal is to reduce Q to a very low value, V_Q. Assume V_Q is low enough such that M4 goes to saturation and M6 goes to linear. After analyzing the write operation with the help of Equation 2.7, we have concluded that a stronger pull-up PMOS and a weaker access device degrades the writability of 6T and RD8T SRAMs as projected in Figure 2.14b. Similarly, in RD8T, as shown in Figure 2.15, the cell enters the write mode by pulling *WWL*, or

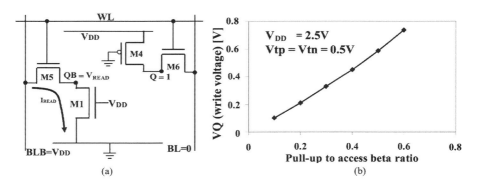

FIGURE 2.14 (a) Write '0' operation of a 6T SRAM cell. (b) Write voltage with respect to the pull-up–access beta ratio.

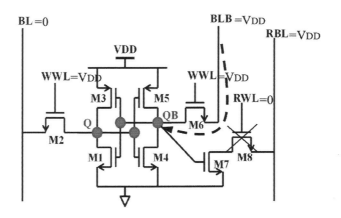

FIGURE 2.15 Write '0' operation of an RD8T SRAM cell.

Design Metrics for Embedded SRAM 55

write wordline, 'high,' and *RWL* remains low, thereby separating the write operation from read, thus eliminating the write operation constraints on the transistor dimensions and improving read margin. Logic stored in write bitline (*WBL*) is written into the storage node Q through M4 and node QB.

2.1.3 LATCH MECHANISM IN SRAM

In this section, we intend to understand what a latch is, what is holding in SRAM, and what factors contribute to data retention in SRAM. So, basically, a latch is a cross-coupled combination of CMOS inverters. Both 6T and 8T have a common latching mechanism since the latch in both the bitcells is formed by two cross-coupled inverters. The transistors M1–M2–M3–M4 in Figure 2.16 represents a latch; a latch has the capability of holding one bit of data until it is provided with a required voltage (V_{min}); when voltage is less than the V_{min} value, the data are lost. Latching is the state when the SRAM cell is idle, and the *BL* and *BLB* are kept at V_{DD} when the access transistors are detached because the word line is not inserted. Thus, the PMOS transistors will continue to strengthen each other if they are connected to the power supply to keep the data. Also, during retention mode, when '1' is stored in the cell, M1 and M4 are ON; hence, there exists positive feedback between the Q and QB nodes, making Q be pulled to V_{DD}. Similarly, when '0' is stored in the cell, M1–M4 are OFF and M2–M3 are ON while QB is pulled to V_{DD}. Now, to keep the data retained, there is minimal voltage to be applied, which is called the data retention voltage (*DRV*) [3]. So *DRV* is not a specifically applied voltage; it is the reduced amount of V_{DD} (V_{min}). In the circuit structure of a 6T SRAM cell, as shown Figure 2.16, when V_{DD} is reduced to *DRV*, all six transistors are in the subthreshold (sub-V_{th}) region, thus the capability of SRAM data retention strongly depends on the sub-V_{th} current conduction behavior.

In a standard SRAM cell, when V_{DD} is reduced to *DRV*, the voltage transfer curves of the internal inverters degrade to such a level that the noise margin (NM) of the

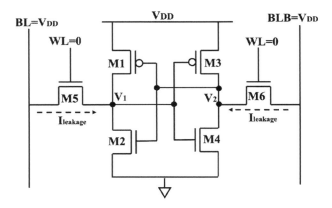

FIGURE 2.16 SRAM leakage current in hold or latch mode [3].

SRAM cell degrades to zero. Using the notations of Figure 2.16, this condition is given by

$$\frac{\partial V_1}{\partial V_2} \text{ (Left inverter)} = \frac{\partial V_1}{\partial V_2} \text{ (Right inverter)}, \text{ when } V_{DD} = DRV$$

If V_{DD} is further reduced below *DRV*, the inverters flip to the state determined by the deteriorated voltage transfer characteristic (VTC) and lose the capability to preserve the stored data. So, after considering the parameters available from Figure 2.16 and solving the sub-V_{th} VTC equations of the two internal data-holding inverters, the *DRV* [3] of an SRAM cell is deduced as

$$DRV = DRV_1 + \left[\frac{V_1}{2} + \left(\frac{DRV_1 - V_2}{2}\right) \cdot n_2\right], \text{ where}$$

$$DRV_1 = \frac{kT/q}{n_2^{-1} + n_3^{-1}} \cdot \log\left[(n_3^{-1} + n_4^{-1})\frac{A_4}{A_2 A_3}\left(\frac{A_5}{n_2} + \frac{A_1}{(n_1^{-1} + n_2^{-1})^{-1}}\right)\right]$$

$$V_1 = \frac{kT}{q} \cdot \frac{A_1 + A_5}{A_2} \cdot \exp\left(\frac{-DRV_1}{n_2 kT/q}\right),$$

$$V_2 = DRV_1 - \frac{kT}{q} \cdot \frac{A_4}{A_3} \cdot \exp\left(\frac{-DRV_1}{n_3 kT/q}\right) \text{ and } kT/q = 26\,\text{mV}$$

The preceding *DRV* formula only relies on the values of A_i (cross-sectional area of the transistors) and n_i (ideality factor), which can be easily extracted from transistor characterizations (either by simulation or by measurement) [3]. The foremost entity affects almost any electronic device or operation is noise. Here we discuss the minimum noise in voltage that can cause the distortion or any change in the storage data by flipping the state of the cell. Typically, this term is defined as the static noise margin (SNM). One of the easiest ways to figure out this noise is by drawing a square of maximum size in the butterfly curve of SRAM as shown in Figure 2.17. If, say, there are two lobes corresponding to this graph and we get two squares from these two lobes, the width of those squares would represent the SNM, and the minimum of the two widths would be the SNM of the SRAM. Figure 2.17 gives an idea of this; however, in the next section, SNM is discussed in detail.

2.2 BITCELL STABILITY METRICS

Cell stability regulates the sensitivity of the memory to process tolerances and different operating conditions. NM is the maximum amount of signal that can be accepted by the device while still maintaining the correct operation. It is assumed that noise

Design Metrics for Embedded SRAM 57

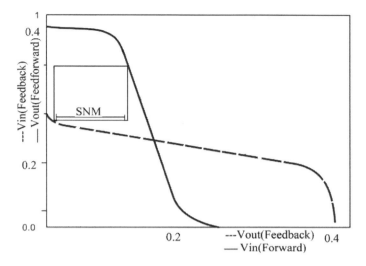

FIGURE 2.17 SNM estimation using the butterfly curve [3].

is present long enough for the circuit to react; that is, the noise is 'static.' Static noise can be DC disturbances such as offsets and mismatches due to processing and variations in operating conditions. An SRAM cell should be designed to work under all conditions, and even some SNM is reserved to tolerate dynamic disturbances caused by particles, crosstalk, voltage supply ripple, and thermal noise. Variations in manufacturing can be classified as systematic or random. Systematic variations are predictable in nature and depend on factors such as layout structure and the surrounding topological atmosphere. However, random variations are unpredictable and are caused by random uncertainties in the fabrication process, such as microscopic fluctuations in the number and location of dopant atoms in the channel region. Random variations are harder to characterize and can have a detrimental effect on the yield of critical modules in a design. Random variations can cause a significant mismatch in neighboring devices. SRAM cells are generally designed to ensure that the contents of the cell do change during the read operation and change states quickly during the write operation. These different read and write requirements are satisfied by altering the relative strengths of the devices in the design. Such careful design of SRAM cells provides stable read and write operations, but it also makes the cells vulnerable to failures caused by random variations in the device strengths. Therefore, an SNM indicates the maximum DC noise amplitudes that can be withstood by the logic. This means that the noise may be present for an infinitely long time without bringing gates to the wrong state. Many analytical models have been proposed to find the SNM of the circuits mathematically as a function of device parameters and supply voltage, to predict the parametric variations on SNM [4–6]. SRAM static stability metrics can be characterized into conventional and large-scale metrics. Conventional metrics are based on a butterfly curve and an N-curve, these metrics require access to internal nodes of the bitcell for measurement of stability. Whereas large-scale metrics do not require access to the internal storage nodes, the measurement of stability is done by accessing the bitlines (*BL* or *BLB*), wordline (*WL*), and supply voltage.

2.2.1 Method of Evaluating NM

There are basically two types of methods used to evaluate the NM in almost all types of SRAM bitcells [4–6]. For evaluating DC NM, the butterfly curve method is widely used, and for dynamic NM, the N-curve method is used [4–6]. There are other types of NM estimation methods available in the literature; however, these two types are very commonly used.

2.2.1.1 Butterfly Curve

A graphical technique called butterfly curves is used to determine the SNM of the cell during the hold, read, and write operations. Figure 2.18 shows the setup for modeling SNM, where DC noise sources with a value V_N are introduced between the inputs and outputs. As V_N increases, the stability of the cell changes. Figure 2.17 shows the butterfly curve for finding SNM graphically for a bitcell holding data. It plots the VTC curve for inverter-L from Figure 2.18 and the inverse VTC from inverter-R, which forms a resulting two-lobed curve called a butterfly curve. SNM is defined as the length of the largest square, which can be inscribed in the lobes of the butterfly curve. It is considered when the value of V_N increases from 0 to V_{DD}. On the plot, this causes the inverse VTC for Inverter 1 in the figure to move downward and the VTC for Inverter 2 to move to the right. Once they both move by the SNM value, the curves meet at two points. After evaluating the SNM, any value more than the SNM can flip the cell. For stability under different operations, different biases for wordlines and bitlines are set. For example, to plot the butterfly curve of hold SNM, wordlines are set to '0,' the potential of a storage node (Q) is swamped from zero to V_{DD}, and the potential of the complementary node (QB) is measured. With this, we can obtain the butterfly curves just by plotting two times, first, by setting Q on the

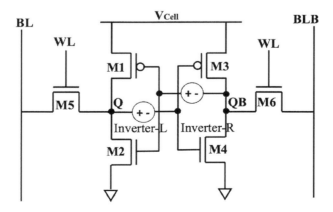

FIGURE 2.18 Experimental setup for extracting a butterfly curve [4].

Design Metrics for Embedded SRAM

X-axis and QB on the Y-axis and, then, superimposing the same plot with the second one by changing the axis (X-QB and Y-Q). But the butterfly curve does not provide current flow information, which is equally vital for stability analysis. The N-curve is used to find the current values as defined in the next section.

2.2.1.2 N-curve

Unlike the butterfly curve, where separate measurements are required for read and write stability and information on current flow dependency on stability is not provided, the N-curve provides both read and write stability metrics at a time. The experimental setup for N-curve measurement is shown in Figure 2.19. The bitlines are precharged, and the wordline is activated for read operation; storage nodes Q and QB are hold '0' and '1,' respectively. Initially, when the voltage source and the storage node Q are at '0,' the access transistor M5 and transistor M2 are in velocity saturation and the linear region, respectively. Therefore, the drain current of M5 is larger than the drain current of M2. To plot the curve, the voltage source is swept at Q from 0 to V_{DD}, and the corresponding current (I) provided by it is also measured. The resulting 'V' and 'I' values are plotted on the X- and Y-axis to form the N-curve. There is no need for mathematical manipulation as the N-curve directly provides the voltage and current stability information of the SRAM cells.

The extracted N-curve as shown in Figure 2.20 has three points A, B, and C; A and C correspond to stable points while B is a metastable point. The N-curve values between A and B define the read stability metrics as shown in the figure: The static voltage noise margin (SVNM) denotes the maximum DC voltage tolerable at the internal node previous to flip the memory cell content, and the static current noise margin (SINM) denotes the maximum DC current value that can be injected in the memory cell before its content changes. Similarly, N-curve values between B and C define the write metrics: the write trip voltage (WTV), as the DC voltage drop needed to flip the memory cell content, and the write trip current (WTI), as the amount of DC current injected in the memory cell to change its content.

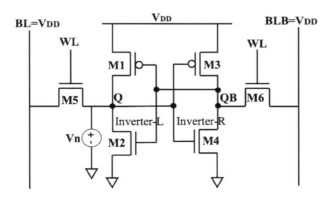

FIGURE 2.19 Experimental setup for extracting N-curve [4].

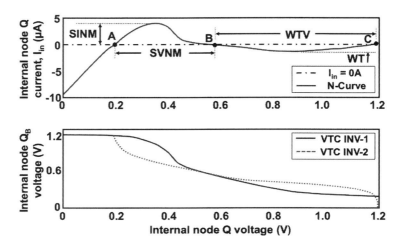

FIGURE 2.20 Extracted N-curve for static current NM and write trip voltage [4–6].

2.2.2 READ STATIC NOISE MARGIN

An SRAM can be prone to failure during a read operation. Transistors must be strong enough to discharge the bit lines without flipping the value stored. An SNM means the maximum amount of voltage noise that can be tolerated at the cross-coupled inverter output nodes or the storage nodes without flipping the cell or the value stored. During the read operation, a low is maintained at one of the storage nodes (Q or QB), with the help of pull-down driver transistors (M2 or M4). Figure 2.21 shows the node values' setup and the noise voltage sources to introduce the disturbance simulating a DC sweep from 0 to V_{DD}. This introduces an extra voltage, if high enough, causes the loss of stored value. The maximum extra voltage tolerated by the cell before losing the data during a read operation is defined as the RSNM. While reading, the cell exposes the storage node (Q, QB) to the disturbance caused by the resistive voltage division between the access and the pull-down devices; that is, these transistors behave like a resistor in this operation and, as a result, form a voltage divider that raises node Q or QB voltage by ΔV or V_{Read}. These disturbances can be minimized by making the pull-down device much stronger than the access transistor. However, random variations in the threshold voltages and the strengths of various devices in an SRAM cell can cause the read operation to flip the contents of the cell. These failures are defined as read stability failures.

From the Equation 2.6 of V_{Read} and V_{trip}, try to maximize the read margin value. This can be done by increasing the trip voltage value and by decreasing the read voltage at the same time. Decreasing the read voltage by using a strong pull-down transistor will, in turn, have a direct impact on the trip voltage, and its value is also lowered. Hence, for a better read margin, a weaker pull-down and stronger pull-up transistor are required (see Figure 2.10), whereas for lowering the read voltage,

Design Metrics for Embedded SRAM   61

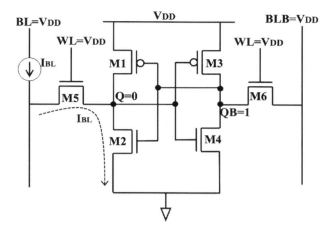

FIGURE 2.21 Schematic diagram for extracting the SRRV [4–6].

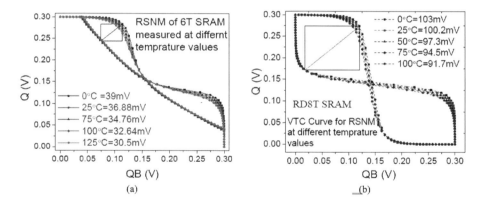

FIGURE 2.22 Butterfly plots for SNM during the read operation for (a) a 6T SRAM and (b) an RD8T SRAM [7].

we need to use weaker access & stronger pull-down transistors. To determine the RSNM, plotting the VTC curve of one inverter and mirroring the VTC of the other inverter during the operation will result in a butterfly curve, and its value is defined as the side length of the largest square that can be fitted inside the lobes as shown in Figure 2.22. A positive value of RSNM represents a stable read operation while a zero or negative value of RSNM signifies that the read operation will cause the cell to lose its state resulting in the read stability failure.

The read stability of a bitcell can also be measured as detailed in the following subsections.

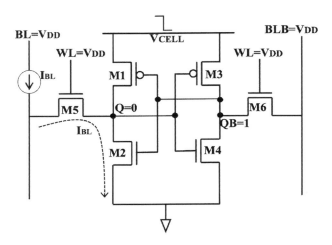

FIGURE 2.23 Setup for extracting SRRV of a 6T SRAM [6].

2.2.2.1 Supply Read Retention Voltage

The minimum bitcell supply voltage is needed for the data retention during a read operation; Figure 2.23 shows the schematic setup for extracting supply read retention voltage (SRVV). The bitline current (I_{BL}) is monitored at the '0' storage node while ramping down the supply voltage of the bitcell. When V_{cell} ramped down sufficiently low, the bitcell loses stability for data retention, thereby flipping the bitcell state. The SRRV of an SRAM bitcell can be defined as the difference between V_{DD} and the value of the V_{cell}, which causes I_{BL} to suddenly drop. If the SRRV is greater than zero, it indicates that bitcell supply voltage (V_{cell}) can be dropped below V_{DD} without disturbing the bitcell state. Therefore, the SRRV represents the maximum tolerable DC noise voltage at the bitcell supply before causing the destructive read operation.

2.2.2.2 Wordline Read Retention Voltage

The largest wordline voltage that can be applied without disturbing the bitcell data retention during the read operation. The experimental setup for finding the wordline read retention voltage (WRRV) is shown in Figure 2.24. However, keeping the wordline voltage below the gate–oxide breakdown voltage set by the technology. The setup for the WRRV is similar to the SRRV except that the wordline here is not fixed at V_{DD}; the cell supply voltage is fixed at V_{DD}. The bitline current (I_{BL}) is monitored at the storage node storing '0' while ramping up the wordline voltage above the supply voltage (V_{DD}). This creates read stress as the access transistor becomes stronger than the driver transistor, so the storage node holding '0' rises above the V_{th} of the other inverter pair and flips the bitcell state. So, the WRRV of an SRAM bitcell is the difference between V_{DD} and the boost in the wordline voltage, causing I_{BL} to suddenly drop. If the WRRV is above zero, it represents that wordline voltage can be boosted above V_{DD} without disturbing the data stored. Therefore, WRRV represents

Design Metrics for Embedded SRAM 63

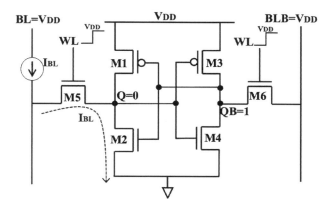

FIGURE 2.24 Schematic diagram for extracting wordline retention voltage [6].

the maximum tolerable DC noise voltage at the wordline before causing the destructive read operation.

2.2.3 WRITE STATIC NOISE MARGIN

An SRAM can be prone to failure during the write operation, and a write operation is performed by setting the bitlines to the required values and enabling access transistors to write the data into both storage nodes (Q, QB). The write SNM can be calculated graphically from a combination of the read and write VTC as shown in Figure 2.25 [7–8]. At one of the storage nodes, QB, voltage is

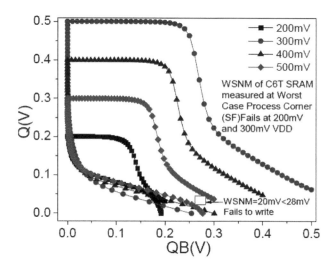

FIGURE 2.25 WSNM estimation for a 6T SRAM [7].

swept from 0 to V_{DD}, and the voltage at the other storage node, Q, is measured for read VTC curve. Both the bit lines are precharged to V_{DD} for this operation. Subsequently, the voltage at storage node Q is swept from 0 to V_{DD}, and voltage at the other node QB is measured for write VTC. Bitlines *BL* and *BLB* are connected to V_{DD} and *GND* for writing '1' and '0,' respectively. The side of the smallest square fitted inside the WVTC and RVTC can be defined as the write static noise margin (WSNM). A larger value of the WSNM ensures a more reliable write operation. The WSNM improves as the access transistor becomes stronger, the pull-up becomes weaker, or the wordline voltage increases. Therefore, efforts are always made to obtain optimum stability during the write operation without affecting the read operation. The strength of the pull-up transistors can be lowered than that of the access transistor, the size of the relationship between the pull-up and access transistors is called the pull-up ratio, and its value is usually lower than 1.

2.2.4 WORDLINE WRITE TRIP VOLTAGE

The stability of the bitcell during a write operation can also be measured externally through wordline. It can be defined as the minimum wordline voltage required to flip the value in the storage node during a write operation, and it can be used to measure the writeability of the SRAM bitcell. To measure the wordline write trip voltage (WWTV) [7–8], the supply voltage and *BL* are pulled high, while *BLB* is pulled low. The bitline storage node voltage holding '0' is monitored while increasing the wordline (*WL*) voltage as shown in Figure 2.26a. As the wordline voltage is ramped high, the storage node voltage Q is monitored. When the voltage of the *WL* is sufficiently high, the bitcell content flips, signified by the sudden drop in the I_{BL} and the crossover point of storage nodes Q and QB as shown in Figure 2.26b. The WWTV is defined as the difference between V_{DD} and *WL* voltage at the point of crossover of Q and QB. The WWTV shows the maximum tolerable drop in *WL* voltage to successfully write into a bitcell.

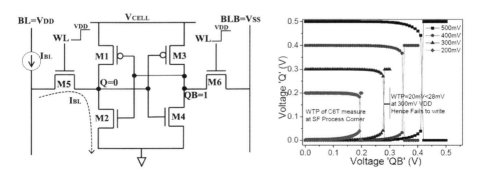

FIGURE 2.26 6T SRAM: (a) Schematic diagram for extracting WWTV. (b) Storage node Q versus V_{WL} plot [7–8].

Design Metrics for Embedded SRAM   65

2.2.5 BITLINE WRITE TRIP VOLTAGE

The write-ability of an SRAM bitcell can also be measured by bitline write margin. The experimental setup for finding bitline write margin is shown in Figure 2.27. The maximum bitline voltage at the storage node QB storing '1,' able to flip during the write operation. To find the bitline write trip voltage (BWTV) of an SRAM bitcell supply voltage, WL and BL are set to V_{DD}, and the bitline current (I_{BL}) is monitored while ramping down BLB voltage from V_{DD}. When the voltage at BLB is dropped low enough and QB reaches below the trip point of the other inverter pair, then a successful write operation is performed, which can be seen as a sudden drop in I_{BL}. If the BWTV is greater than zero, it indicates that the voltage at BLB can be dropped below V_{DD} for a successful write operation. So the BWTV estimates the cell writability as the maximum bitline voltage tolerated by the BLB node able to flip the cell value during a write cycle.

2.2.6 DYNAMIC READ STABILITY

Under the static condition, the SRAM cell wordline (WL) is turned ON indefinitely, and the bitlines (BL and BLB) are driven to V_{DD} indefinitely. Compared to dynamic effects, such as finite duration of WL pulse and floating bitline voltage after precharging, these SNMs can be pessimistic in the case of read stability. In the transient operation of the bitcell, the wordline is a finite pulse, which means storage nodes are not infinitely susceptible to flip. Bitlines are initially precharged and left floating, getting discharged by the bitcells in the array. Some dynamic effects include the following [9–11]:

1. **WL opening:** When a WL is turned on, a sudden change from '0' to '1' can create some instability due to the capacitive coupling between the WL signal at the gate of the access transistors (M5) and the internal

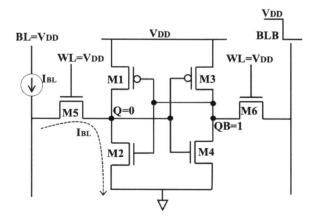

FIGURE 2.27 Schematic diagram for extracting the BWTV for a 6T SRAM.

storage node (Q). The amount of coupling depends on the capacitances of both.

2. **Finite *WL* pulse:** During the T_{on} cycle, the access transistors are turned on, providing bitlines access to the internal nodes. The longer the pulse width (PW) of T_{on}, the worse will be the read stability since it gives more opportunity for the node to flip its stored data. Beyond a certain PW, the dynamic NM saturates and becomes closer to the SNM. The smaller the PW, the higher will be the NM, and again, PW cannot be shrunk to a point where the SA cannot detect the correct stored value due to not reaching the offset voltage, which might result in a read failure as shown in Figure 2.28, the dynamic NM for different pulses.

3. **Bitline discharge:** Once the bitlines are precharged, they are left floating until the wordline is activated; besides the cell being read, other unselected cells in the same column may draw leakage currents from either the bit or bit-bar lines depending on the stored value, and increasing the number of cells in an array degrades the dynamic read noise margin (DRNM) as parasitic capacitance increases. Eventually, dynamic noise margin converges to SNM. It should be noted that dynamic behavior is independent of the WL PW while it mainly depends on the cell parasitic capacitances. If the time between successive read operations on a cell is reduced then there can be an increase in noise from the previous and the current operation that can lead to bit-flip and read failure.

Considering the DRNM, the target noise margin can be achieved with a lower cell ratio compared to the SNM, and the addition of small capacitance at the storage node will improve the stability.

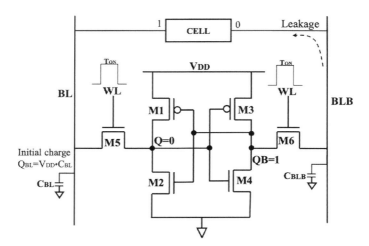

FIGURE 2.28 Read DNM circuit setup for a 6T SRAM.

2.2.7 Dynamic Write Stability

SNM measurements often underestimate write failures. Accurate analysis can be done in a transient situation. During a write '1' operation, the *BL* is charged to V_{DD}, and *BLB* is pulled low as shown in Figure 2.29. When the *WL* is activated, if the *WL* PW is too low, then the data in the storage nodes (Q, QB) may not be overwritten, and they go back to the previous states. The dynamic write failure, therefore, can be defined as if the voltage on the node initially storing '1' (QB) could not reach the trip point of the other inverter during the *WL* pulse, then the bitcell will not be successfully written. So a sufficient *WL* pulse is required to perform a successful dynamic write operation.

2.3 DECODERS

The decoders are the key components for addressing the memory bitcell location in the SRAM array. The whole SRAM array is divided into a three-dimensional matrix of rows, columns, and blocks. There are three decoders used to access the bitcell as shown in Figure 1.2: one column decoder (Z-Decoder), which selects the block; one column decoder (Y-Decoder), which selects the column of the array; and one row decoder (X-Decoder), which selects the row of the array. For example, if we have 512 row locations in an array, then an M-to-2^M Decoder is required to access the bitcells, where M = 9. In a row or column decoder, the collection of 2^M complex logic gates organized in regular and dense fashion. There are two ways a decoder can be designed. We can either use an NAND-based or an NOR-based decoder as shown in the following expressions:

WL expression for NAND Decoder:

$$WL_0 = A_0 A_1 A_2 A_3 A_4 A_5 A_6 A_7 A_8 A_9$$

$$WL_{511} = \overline{A_0} A_1 A_2 A_3 A_4 A_5 A_6 A_7 A_8 A_9$$

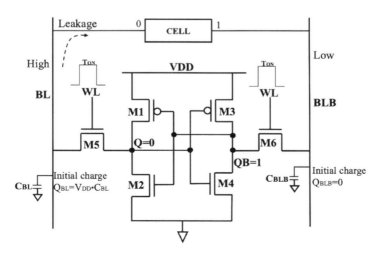

FIGURE 2.29 Write DNM circuit setup for 6T SRAM.

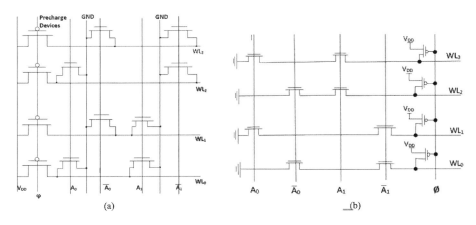

FIGURE 2.30 2-input decoder circuit: (a) NOR-based decoder and (b) NAND-based decoder.

WL expression for NOR Decoder:

$$WL_0 = \overline{A_0 + A_1 + A_2 + A_3 + A_4 + A_5 + A_6 + A_7 + A_8 + A_9}$$

$$WL_{511} = \overline{A_0 + \overline{A_1} + \overline{A_2} + \overline{A_3} + \overline{A_4} + \overline{A_5} + \overline{A_6} + \overline{A_7} + \overline{A_8} + \overline{A_9}}$$

Both NAND and NOR decoders use the same numbers of logic gates, although due to fast switching speed in NAND gates, NAND decoders are widely used for high-speed memory computing. There are dynamic decoders that can also be used for SRAM addressing. The two types of dynamic decoders shown in Figures 2.30a and Figure 2.30b as an examples of size 2-input NOR and 2-input NAND decoders, respectively. The switching MOS devices can be replaced by a transmission gate logic (TGL) switch to design a 2-input NOR decoder shown in Figure 2.31.

2.4 READ DRIVER

The read driver, also known as an SA, is used to amplify a small analog differential voltage developed on the bitlines during a read access cycle. The amplification results in a full-swing single-ended digital output. Employing an SA reduces the size of the SRAM cell as the driver transistors do not need to fully discharge the bitlines. For proper functionality of an SA, it needs to satisfy a few electrical requirements. First, the minimum differential voltage swing required at the input of the SA should be smaller than the minimum differential voltage that is developed over the bitlines by the SRAM cell. Second, the SA should be able to provide the output within the sense amplification time once it gets a minimum input differential voltage. In this book, for sensing data from SRAM, a current latch SA (CLSA) is used.

The CLSA shown in Figure 2.32 is a bilaterally symmetric circuit with the inverters of each branch (MP1–MN1 and MP2–MN2) cross-coupled to form a

Design Metrics for Embedded SRAM

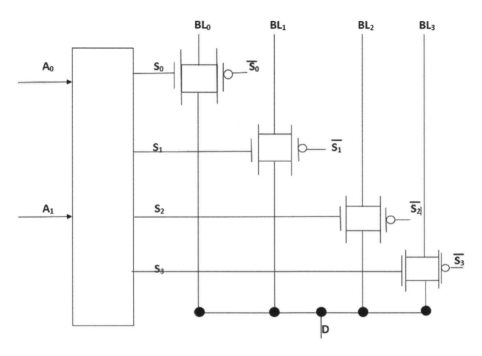

FIGURE 2.31 4-input transmission gate (TG)-based column decoder/multiplexer (MUX).

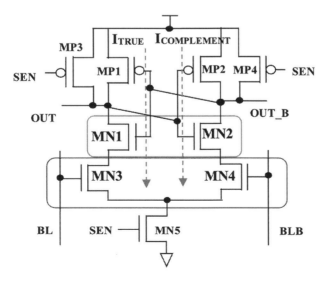

FIGURE 2.32 CLSA circuit.

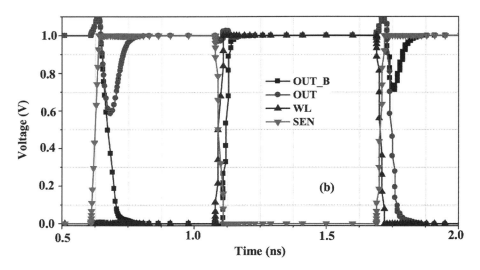

FIGURE 2.33 Read operation using CLSA. *SEN* is the enable input to the SA, and *OUT/OUT_B* are the outputs of the SA.

latch. The input transistors (MN3 and MN4) receive input on their gate from the two bitlines. The data on the memory cell develops a lower voltage on one of the two bitlines compared to the other; thus, the two input transistors have different gate-to-source voltages. As a result, different currents flow through the two transistors, which results in slower discharge of one output node (*OUT* or *OUT_B*) compared to the other. Ultimately, the positive feedback mechanism provided by the cross-coupling of the inverters latches the slower discharging output node to V_{DD} and the other node to *GND*. The symmetry of the CLSA circuit is of critical importance for reliable read operation in SRAMs, which employ a differential sensing scheme. A mismatch in the corresponding devices of the two branches may neutralize or even reverse the effect of inputs and cause a read failure. For example, if, the data to be read is '1' but the V_{th} of MN3 is higher than the V_{th} of MN4, then even though the gate to source voltage of MN3 is higher than that of MN4, the current in MN4 might still be greater than MN3 due to an overall higher overdrive voltage. This will result in the CLSA sensing the data as 0 as shown in Figure 2.33.

2.5 WRITE DRIVER

A write driver, shown in Figure 2.34, is responsible for discharging one of the bitlines from the precharge level to a level below the write margin of the cell before or while the wordlines of the selected cell are active. Normally, the write driver is enabled by the write enable (*WE*) signal and drives the bitline using full-swing discharge from the precharge level to ground. The order in which the wordline is enabled and the write drivers are activated is not crucial for the correct write operation. In a write

Design Metrics for Embedded SRAM 71

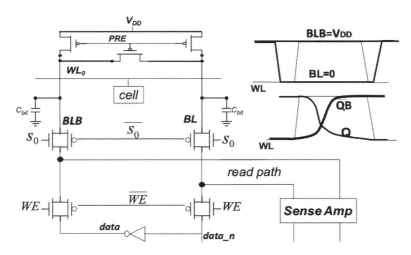

FIGURE 2.34 Write driver for SRAM.

operation, the *WE* signal enables the write path, and the SA-enable signal disconnects the *BL* and *BLB* from the read path. S_0 signal is for connecting the write driver and read driver to the *BL* and *BLB* of the SRAM array.

2.6 LEAKAGE POWER IN SRAM

As an SRAM's demand grows, the density of the SRAM increases. This is achieved by reducing the size of the memory cell, which significantly increases the leakage power. Applications such as Internet of Things (IoT) vary extensively as do the purposes of SRAM in these devices. Power leakage is one of the greatest challenges that threaten to stop the exponential growth of transistors along with process variation. As the size of transistors scales down, the leakage current increases exponentially. Scaling down the transistor also decreases reliability exponentially. These effects induced by scaling down are further aggravated in energy-efficient low-voltage devices. It can be seen from Figure 2.35 that the SNM of a 6T bitcell is gradually decreasing with technology scaling, while the leakage current is increasing exponentially. Moving from 130-nm to 32-nm technology nodes, there is a 55% reduction in the read SNM, but there is an 86% increase in leakage current. But in more recent years, we have reached transistor sizes as small as 5 nm even in consumer electronics. The 7-nm and 5nm technology node devices are affected by two reasons: First, due to the increase in dynamic current density, the overall device heat dissipation increases; second, various types of leakage currents like subthreshold leakage, gate leakage, and junction leakage also rise at the lower technology nodes as shown in Figure 2.36a.

Traditional solutions to leakage power are dependent on architectural level changes or clock gating methods. Clock gating is a technique used to reduce power leakage by removing the clock signal when it is not in use. It can help lower the

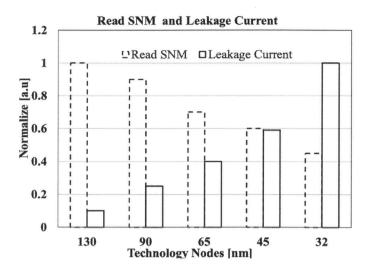

FIGURE 2.35 The comparison of normalized RSNM and leakage current of a 6T SRAM bitcell for different technology nodes [12].

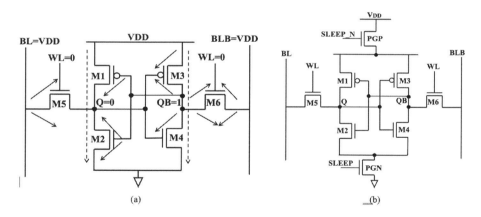

FIGURE 2.36 (a) Leakage currents in a 6T SRAM cell [13]. (b) Schematic of the power-gated SRAM bitcell.

power dissipation and the size of the SRAM cell, which can lead to lower average power consumption of the system. Scaling CMOS technology nodes increases the static power dissipation. This constitutes of the subthreshold leakage current, gate leakage current, and junction leakage current. Figure 2.36b represents an example of a power-gated SRAM memory cell. By shutting off the power supply in the standby state, leakage power is greatly reduced. Power-gated SRAM cells come with their own drawbacks, such as slow switching speeds and more complex architecture.

There are also power-gated SRAM cells with a header switch and a footer switch that have different use cases. A footer-switch SRAM uses a high-V_{th} n-type metal-oxide semiconductor (NMOS) footer switch that is situated between the physical ground and the virtual ground. Footer-switch implementation has better performance but has more power leakage than a header switch. A header-switch SRAM uses a high-V_{th} PMOS transistor embedded between the physical power supply and the virtual power rail. Header switch implementation has lower leakage but worse performance. Body biasing is another technique that helps dissipate thermal and power leakage. Subthreshold leakage is the biggest power consumer in many high-performance designs. Body biasing allows us to dynamically adjust the threshold voltage of a CMOS transistor. Normally, a CMOS transistor has only three terminals, but it is not common to see a fourth terminal at the body. This can be thought of as a second gate, allowing us to control how the transistor turns off and turns on.

2.7 DYNAMIC POWER IN SRAM

The read and write dynamic power consumption of an SRAM unit is often calculated by adding up the dynamic power consumption related to different capacitive loads that are charged and discharged during the read and write operations, respectively. Clearly, the entire dynamic power consumption is especially dominated by the long interconnects, which impose an outsized capacitive load to the signal paths within the SRAM unit. For example, the outputs of the pre-decoders at the row decoder, which are loaded by the post-decoder; the block decoder's output; the wordlines and bitlines; and the column decoder's output are considered as heavily loaded interconnects. The dynamic power consumption of the SRAM unit is especially important when the speed of operation is high. The types of dynamic power consumption are mentioned in Section 1.3.1.1.

2.8 READ/WRITE LATENCY

It is the time taken for the I/O request to be completed. The time measurement is started when the moment request is issued to the storage layer and stops measuring when either we get the required data or we get the confirmation that the data are stored on the disk. Latency is the single-most important metric to focus on when it comes to storage performance. The access time of the unit is the sum of the latency of several delay components.

$$Total\ Time = T_{Buff} + T_{Decoder} + T_{Bitline} + T_{SenseAmp} + T_{DataBus} + T_{OutBuff}$$

T_{Buff} is the delay time of the input address buffer, $T_{Decoder}$ is the delay of the decoder, $T_{Bitline}$ is the time needed for the discharge of the bitlines, $T_{SenseAmp}$ is the delay of the sense amplifier, $T_{DataBus}$ is the delay of the data bus, and $T_{OutBuff}$ is the delay of the output latches. The self-aligned contact (SAC) [14] scheme and the segmented virtual grounding (SVGND) [15] only influence the $T_{Bitline}$ compared to the conventional scheme. Hence for the same memory size with the same peripheral circuits, the rest of the time delays remain the same for all three schemes including conventional.

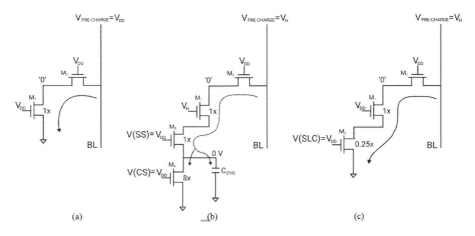

FIGURE 2.37 (a) Conventional SRAM, (b) SVGND scheme [15], and (c) SAC scheme [14].

$T_{Bitline}$ increases in SAC and SVGND compared to the conventional scheme when the same transistor sizes are used because of the extra switches in the bitline discharge path for a constant bitline capacitive load. Figure 2.37 shows the bitline discharge path of the conventional scheme, SAC, and SVGND during the read operation. In this figure, the bitline that is discharged is shown since this is the only side that communicates the data over the bitline. M1 is the drive transistor, and M2 is an access transistor. Assuming the same SA is used, the bitline must be discharged by the same voltage in all three schemes. In SVGND, the column virtual ground (CVG) is discharged to the ground through M4. M4 is designed to be a large transistor and is shared for the whole column. In SAC, the same virtual ground switch (i.e., same M3 size) is shared among four cells in a row, hence four transistors are discharged through this transistor. Therefore, the effective W/L associated with one cell is one fourth. CVG and M3 are the switches that connect the virtual ground of the cell to the ground in SVGND and SAC schemes.

REFERENCES

1. Pavlov, A., and Sachdev, M. (2008). *CMOS SRAM Circuit Design and Parametric Test in Nano-Scaled Technologies*. Springer. https://doi.org/10.1007/978-1-4020-8363-1.
2. Wuu, J., Weiss, D., Morganti, C., and Dreesen, M. (2005). The asynchrounous 24MB on-chip level-3 cache for a dual-core Itanium-family processor. *IEEE International Solid-State Circuits Conference*, pp. 488–489.
3. Qin, H., Cao, Y., Markovic, D., Vladimirescu, A., and Rabaey, J (2004). SRAM leakage suppression by minimizing standby supply voltage, *Department of EECS*, University of California at Berkeley. https://dl.acm.org/doi/10.5555/977402.978177
4. Seevinck, E., List, F. J., and Lohstroh, J. (1987). Static-noise margin analysis of MOS SRAM cells. *IEEE Journal of Solid-State Circuits*, 22(5), pp. 748–754. https://doi.org/10.1109/jssc.1987.1052809.
5. Lohstroh, J., Seevinck, E., and de Groot, J. (1983). Worst-case static noise margin criteria for logic circuits and their mathematical equivalence. *IEEE Journal of Solid-State Circuits*, 18(6), pp. 803–807. https://doi.org/10.1109/jssc.1983.1052035.

6. Calhoun, B. H., and Chandrakasan, A. P. (2006). Static noise margin variation for sub-threshold SRAM in 65-nm CMOS. *IEEE Journal of Solid-State Circuits*, 41(7), pp. 1673–1679. https://doi.org/10.1109/jssc.2006.873215.
7. Singh, P., and Vishvakarma. S. K. (2017). Ultra-low power, process-tolerant 10T (PT10T) SRAM with improved read/write ability for internet of things (IoT) applications. *Journal of Low Power Electronics and Applications*, 7(3), p. 24. https://doi.org/10.3390/jlpea7030024.
8. Ruchi, R., and Dasgupta, S. (2017). Compact analytical model to extract write static noise margin (WSNM) for SRAM cell at 45-nm and 65-nm nodes. *IEEE Transactions on Semiconductor Manufacturing*, 31(1), pp. 136–143. http://doi.org/10.1109/tsm.2017.2772341.
9. Elthakeb, A. T., Haine, T., Flandre, D., Ismail, Y., Elhamid, H. A., and Bol, D. (2015). Analysis and optimization for dynamic read stability in 28nm SRAM bitcells. *2015 IEEE International Symposium on Circuits and Systems (ISCAS)*. http://doi.org/10.1109/iscas.2015.7168908.
10. Dong, W., Li, P., and Huang, G. M. (2008). SRAM dynamic stability: Theory, variability and analysis. *2008 IEEE/ACM International Conference on Computer-Aided Design*. http://doi.org/10.1109/iccad.2008.4681601.
11. Toh, S. O., Guo, Z., Liu, T.-J. K., and Nikolic, B. (2011). Characterization of dynamic SRAM stability in 45 nm CMOS. *IEEE Journal of Solid-State Circuits*, 46(11), pp. 2702–2712. http://doi.org/10.1109/jssc.2011.2164300.
12. Singh, J., and Vijaykrishnan, N. (2013). A highly reliable NBTI resilient 6T SRAM cell. *Microelectronics Reliability*, 53(4), pp. 565–572. https://doi.org/10.1016/j.microrel.2012.11.003.
13. Zhang, Li-Jun, Wu, Chen, Ma, Ya-Qi, Zheng, Jian-Bin, and Mao, Ling-Feng. (2011). Leakage power reduction techniques of 55 nm SRAM cells. *IETE Technical Review*, 28, pp. 135–145.
14. Kanda, K., Hattori, S., and Sakurai, T. (2004). 90 % write power-saving SRAM using sense-amplifying memory cell. *IEEE Journal of Solid-State Circuits*, 93, pp. 929–933.
15. Sharifkhani, M., *Design and Analysis of Low-Power SRAMs*. UWSpace. http://hdl.handle.net/10012/2870.

3 SRAM Bitcells over Conventional Memories

3.1 INTRODUCTION

Over the past few decades, memory has been an important component behind the fast growth of complementary metal-oxide semiconductor (CMOS) technology. The influx of virtual memory contributed to the hierarchical structure of several kinds of memory in computers and handheld electronic devices. The types of memory vary from lesser capacity to more costly but fast cache memories to more capacity, affordable but slow magnetic and optical storage. Moore's law of scaling plays an important role in technological advancement in the semiconductor industries for many decades. The drive to meet the requirements of Moore's law causes a shrinking of the thickness (W) and channel length (L) of the transistor gate. Additionally, the scaling reduces the supply voltage and power needed for reliable operation with improved functionality and performance [1,2]. This trio of functionality, higher performance, and low power requirement provides direct benefits to consumers, who continuously require low-energy on-chip embedded memory. Moreover, according to the semiconductor industry, embedded static random-access memory (SRAM) has considered as a critical technology enabler for a wide range of applications such as high-performance computing, wearable electronics, the Internet of Things (IoT), and image-processing systems as compared to dynamic random-access memory (DRAM) and synchronous dynamic random-access memory (SDRAM) [3]. This is due to the highest write bandwidth and read latency of SRAM with low-power operation as demonstrated in Table 1.1, which shows the performance metrics of state-of-the-art memory technologies. To get effective benefits from transistor scaling, modern digital systems stress the use of more and more SRAMs [4,5]. The subsequent consequence of modern systems on chip (SoCs) is that embedded SRAM covers a significant portion of SoCs and has a large impact on the performance of the chip. Since the portion of SRAM in SoCs is increasing every year, it is predicted to cover about 80%–90% of the die area by 2020 [6,7].

Moreover, the power share of memory in Intel Core i7-6700k and ARM Cortex A72 processor is 40%–50% [8,9]. The most crucial trend observed here is that the SRAM power becomes more significant since it occupies a core portion of the total die area and accounts for the largest share of power consumption in a system. In SRAM, power reduction can be attained by the subthreshold operation, which reduces the robustness due to lower noise margins and higher susceptibility to process variations. Another challenge is statistical process variations in the transistor parameters such as threshold voltage (V_{th}), channel length (L), and mobility [10,11]. Thus, device variability in modern processes of SRAM design has become a major concern, which degrades the performance, bit density, power, reliability, and yield.

Low-voltage operation with increased process variation caused by random dopant fluctuation (RDF) and line edge roughness (LER) degrades the stability and performance of SRAM, which may lead to functional failure [12]. Therefore, there are basically four key traits to consider in the subthreshold SRAM architectures namely, standby power, read-write energy, read-write cell stability and read failure. To overcome, all these factors along with low standby power and better cell stability, differential/single-ended ultra-low-power process-tolerant SRAM architectures are presented in this chapter. The SRAMs presented in this chapter have come up with improvements in parameters that can be associated with ultra-low-power application devices such as handheld devices, medical equipment, field programmable gate arrays (FPGAs), image processing, and IoT edged devices.

3.2 SRAM IN SUBTHRESHOLD FPGA

It is observed that the leakage power and cell stability are the key constraints in the subthreshold SRAM-based FPGA to improve reliability, yield, and susceptibility of portable electronic devices. SRAM remains in a static or hold state for most of the time, which contributes to more leakage power. Moreover, portable electronic devices have extremely low power requirements to increase the battery lifetime. Furthermore, the noise generated from threshold variation, process variation, and multiple-bit errors reduces the stability of SRAM cells. Consequently, various techniques have been employed to overcome these limitations, such as scaling the supply voltage using process variation tolerant Schmitt trigger (ST)–based 10T SRAM [9] shown in Figure 3.1e, a read static noise margin free 7T SRAM [10], differential data-aware power supplied D2AP8T SRAM [11] shown in Figure 3.1c, and a low leakage variation tolerant LP8T SRAM cell for ultra-low power applications [12] shown in Figure 3.1d. Besides the advantages provided by these cells, there are limitations like read stability, consequences from temperature variations in leakage power, low write trip point (WTP), and higher power delay product (PDP) in the subthreshold region. To compensate for all these factors along with low leakage power and better cell stability, a positive feedback controlled 10T (PFC10T) SRAM is illustrated here.

In the literature, there are reported works on feedback cutting of SRAM cells [13,14]. In Chang et al.'s study [13], feedback cutting is used for read–write operations to improve read and write stability. Similarly, in [14], 9T SRAM cell with a data-aware feedback-cutoff (DAFC) scheme is projected to improve the write margin and dynamic-read-decoupled (DRD) scheme is used to prevent read-disturb in the subthreshold operations. The main limitation in [13] is due to a single-ended read operation, which adds a negative voltage bump at the storage node while reading and limits the value of the read static noise margin (RSNM). Furthermore, the single-ended write slows down the write process and limits the value of write static noise margin (WSNM) and WTP. Subsequently, it is also observed that feedback controlling increases the layout area of the SRAM array. However, it improves other parameters such as cell stability, leakage power, and read–write energy at different process, voltage, and temperature (PVT) values.

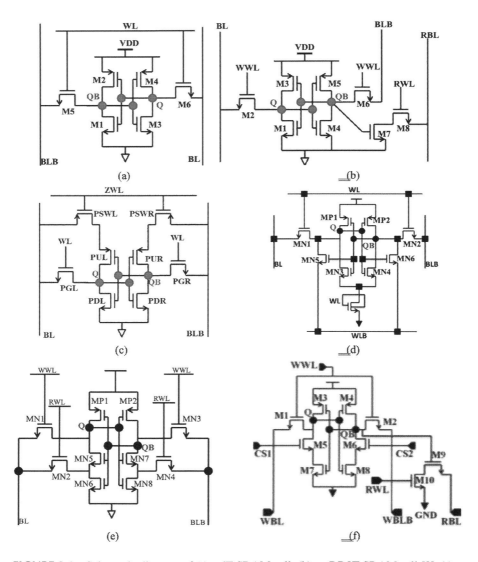

FIGURE 3.1 Schematic diagram of (a) a 6T SRAM cell, (b) an RD8T SRAM cell [8], (c) a D2AP8T SRAM cell [11], (d) an LP8T SRAM cell [12], (e) an ST10T SRAM cell [9], and (f) a read-decoupled PFC10T SRAM.

3.3 PFC10T FOR LOOK-UP TABLE

Cell stability and leakage power are considered major concerns in subthreshold SRAM–based FPGA devices. Typically, the bitcell used in a look-up table (LUT) of FPGA is conventional 6T (C6T) SRAM. However, the conventional SRAM cell has failed to achieve better cell stability and low leakage power in the subthreshold

SRAM Bitcells over Conventional Memories

region [15,16]. It is also observed that RSNM is vulnerable at the subthreshold regions while reading through a full-swing sense amplifier (SA) [17,18]. As a result, to work in the subthreshold region with greater stability and robustness, researchers have developed nonconventional SRAM architectures [2–8], such as 8T SRAM cells that utilize reverse short channel effects for increased write-ability at lower voltages [5,17] and an 8T SRAM for variability tolerance and low-voltage operation in high-performance FPGAs and caches [8]. These memory architectures have achieved tremendous improvements in the performance of SRAM cells in terms of cell stability and speed. However, the leakage power is still a concern in the subthreshold FPGA operations. As a result, a read-decoupled (RD) 10T SRAM cell is demonstrated here to overcome the leakage power issue with better cell stability. The featured cell achieves better cell stability in terms of RSNM, WSNM, and dynamic read margin (DRM). The cell is further used as a storage device in a 6-input LUT and a 2-kilo bit (2-kb) SRAM array.

3.3.1 ARCHITECTURE OF PFC10T SRAM

A PFC10T SRAM cell using United Microelectronics Corporation (UMC) 65nm standard CMOS technology is presented. The PFC10T works on the principle of feedback cutting using two control inputs namely *CS1* and *CS2* as shown in Figure 3.1f. The layout of a PFC10T SRAM cell is shown in Figure 3.2d. The cell layout area of the cell is compared with a C6T SRAM cell, an RD-8T SRAM cell, a D2AP8T SRAM cell, and an ST10T SRAM cell as shown in Table 3.1. It shows that

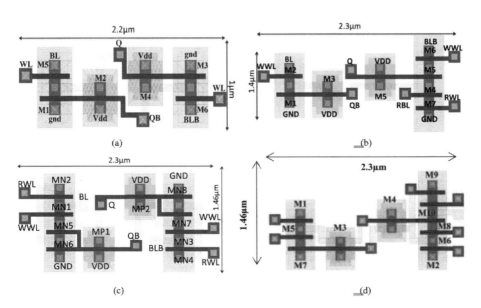

FIGURE 3.2 Layout of (a) a 6T SRAM, (b) an RD8T SRAM [8], (c) an ST10T SRAM [9], and (d) a PFC10T SRAM.

TABLE 3.1
Comparison of Cell Layout Area

SRAM cell	Number of transistors used	Normalized layout area
6T	6 (2 LVT PMOS, 4 LVT NMOS)	1×
RD8T [8]	8 (2 LVT PMOS, 6 LVT NMOS)	1.45×
D2AP8T [11]	8 (2 LVT PMOS, 6 LVT NMOS)	1.35×
ST10T [9]	10 (2 LVT PMOS, 8 LVT NMOS)	1.45×
PFC10T	10 (2 LVT PMOS, 8 LVT NMOS)	1.45×

the featured cell occupies 45% more footprint area as compared to C6T SRAM and has a similar layout area compared to RD8T and ST10T SRAM cells. Nevertheless, a significant improvement is attained in static noise margins and leakage power at different PVT conditions in subthreshold supply voltages by using feedback cutting.

3.3.1.1 PFC10T SRAM Bitcell

The PFC10T cell (refer to Figure 3.1f) consists of two write access n-type metal-oxide semiconductor (NMOS) transistors, M1 and M2. An input bit is written to the SRAM cell through these two transistors. The *WBL* contains single-bit information to write, and at the same time, *WBLB* contains the complementary of that. However, NMOS transistors M9 and M10 are used as a read decoupled logic to read information from the cell. The *RBL* is charged to V_{DD} before the read operation is performed. The M3 to M8 transistors form a latch, where M3 and M4 are the pull-up p-type metal-oxide semiconductor (PMOS) transistors connected to the V_{DD}. In addition, M7 and M8 are the pull-down transistors connected with *GND*. Additionally, M5 and M6 are the feedback-cutting transistors used to disconnect the path between V_{DD} to *GND*. The positive voltage feedback cutting of the SRAM latch is controlled through two NMOS transistors M5 and M6. These two transistors are activated by control signals *CS1* and *CS2* shown in Figure 3.3, which lead to an improvement in the read-write speed, SNMs, and the leakage power of the cell. Operations at different states are explained in Table 3.2. The controller block for the 10T SRAM cell and the SRAM array are detailed in further subsections. The controller shown in Figure 3.3 plays a major role in generating *CS1* and *CS2* signals corresponding to the read-write-hold states.

3.3.2 SRAM WITH FEEDBACK CONTROL CIRCUIT

The control signals *CS1* and *CS2* are provided by a 32-bit bus architecture, which is shared among the 32-bit SRAM cells of a column in a 32 × 64 (2-kb) SRAM array as shown in Figure 3.4a. The control signals *CS1* and *CS2* remain in the previous state after the write operation is performed. The state of control signals *CS1* and *CS2* only gets changed after the next write operation. Therefore, the states of *CS1* and the states of *CS2* are identical in both hold and read operations, for instance, it

SRAM Bitcells over Conventional Memories

TABLE 3.2
Truth Table for Various Operations in a PFC10T SRAM

Operation	WWL	RWL	WBL	WBLB	RBL	CS1	CS2
Write '1'	1	0	1	0	0	0	1
Write '0'	1	0	0	1	0	1	0
Read '1'	0	1	0	0	1	0	1
Read '0'	0	1	0	0	Discharge	1	0
Hold '1'	0	0	0	0	1	0	1
Hold '0'	0	0	0	0	1	1	0

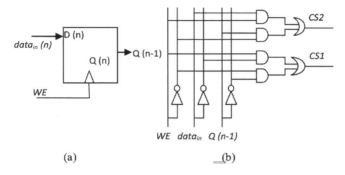

FIGURE 3.3 Circuit of a controller block: (a) Latch to store the input data ($data_{in}$) and (b) the control signal generator for a PFC10T SRAM.

is observed from Table 3.2 that in read '1' and hold '1' operations, the state of *CS1* is at logic '0' and the state of *CS2* is at logic '1'. Since it is the most significant condition to consider, a controller, shown in Figure 3.4b, is placed at every column of the SRAM array. Figure 3.4b shows the read-write control circuitry of controller logic for an SRAM.

The bitlines *WBL, WBLB* and *RBL* carry single-bit information, which is selected by the read-write circuitry (consisting of a row and column address decoder). In the case of the write operation, the combination of *WWL* (controlled by the row address decoder) and bitlines (controlled by column address decoder) selects the SRAM cell from the SRAM matrix. After enabling the write operation of an SRAM cell from row and column decoders, the *WBL* writes bit information into the bitcell. The read operation is also performed in a similar manner, whereby the column address decoder selects the particular bitline to read, and the *RWL* signal (which is enabled by the row address decoder) selects the bitcell from the row. The SA senses the voltage difference between the *RBL* and power supply (V_{DD}) line. After attaining the required sensing voltage, the SA detects the actual bit stored in the SRAM bitcell, and further, the read-write circuitry reads the output as logic 1 or logic 0. The control signals *CS1* and *CS2* are used to hold the state

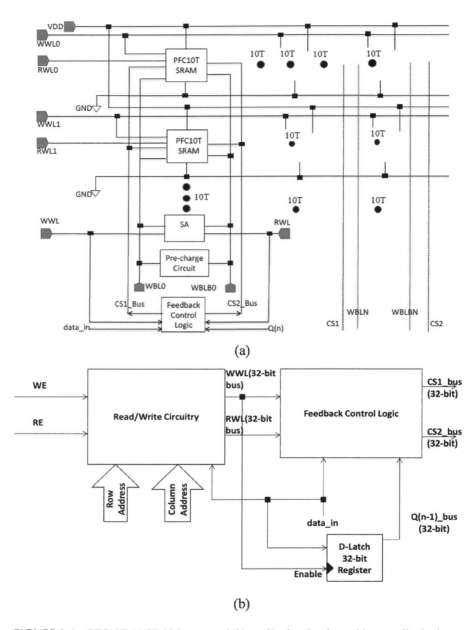

FIGURE 3.4 PFC10T (a) SRAM array and (b) read/write circuitry with controller logic.

of Q and QB in standby and read mode. This is only possible by maintaining the states of *CS1* and *CS2* similar to that of the write operation. For instance, if a write '1' operation is performed, then *WWL* and *RWL* are kept HIGH and LOW, respectively. The data input $data_{in}$ is written using the read/write circuitry as shown in Figure 3.4. The row address decoder and the *WE* input select the row from the SRAM array by enabling *WWL* HIGH. The column address selects the *BL* and *BLB* to write input data $data_{in}$. At the same time, the *WWL*, $data_{in}$, and $Q(n-1)$ determine the state of the *CS1* and *CS2* buses. The *RWL* and the *WWL* are shared with 64-bit SRAM cells in each row of the 32 × 64 SRAM array. It is noted that the D-latch-based 32-bit register is used to store the previous information of input data when *WWL* is LOW. Therefore, $Q(n-1)$ 32-bit output of D-latch retains the state of *CS1* and *CS2* 32-bit buses when *WWL* is disabled.

3.3.3 READ OPERATION IN PFC10T SRAM

The read operation in the 10T SRAM is explained in Figure 3.5a. Before reading information from *RBL*, it is charged to V_{DD}. The read operation is obtained through ultra-fast current mode SA [19]. The read operation is performed by keeping *RWL* to HIGH and *WWL* to LOW. For a read '0' operation, the control signals *CS1* and *CS2* are kept at HIGH and LOW, respectively. For read '0' (Q=0), logic 1 is stored at storage node QB and *RWL* is kept HIGH, which would turn ON M9 and M10, respectively. *RWL* turns M10 ON, and a discharging path from RBL-M9-M10-GND is formed.

The voltage difference, $\Delta V_{RBL} = \{V_{DD} - (V_{DD} - I_{read} \times R_{M9-10})\}$ appears between *RBL* and a V_{DD} line, which is sensed by the full-swing inverter SA. In the preceding equation, I_{read} is the cell current and R_{M9-M10} is the resistance through M9 and M10. Figure 3.5a shows a discharging path from *RBL* to *GND* for a read '0' operation. The read time is measured as the time from the *RWL* signal is activated until the *RBL* is discharged to the minimum required potential needed by the SA to read [19].

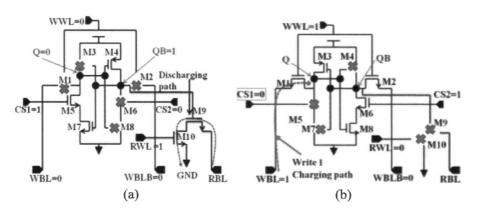

FIGURE 3.5 Schematic of PFC10T cell (a) in a read operation and (b) in a write '1' operation.

3.3.4 Write Operation in PFC10T SRAM

Figure 3.5b shows the write '1' operation of PFC10T SRAM. For a write operation, *WWL* must be kept HIGH and the *RWL* kept LOW. To write logic 1, the control signals *CS1* is kept LOW and *CS2* is kept HIGH. Thus, by keeping *CS1* LOW, transistor M5 turns OFF, which would disconnect the discharging path through Q. As a result, the storage node Q would charge fast through *WBL*. Similarly, at storage node QB, M6 turns ON by asserting *CS2* HIGH, which makes the discharging path across QB to *GND*. Therefore, information stored at QB would discharge quickly, which makes write '0' fast. The write '1' time is measured as the time from when *WWL* signal is activated and storage node Q reaches 90% of the V_{DD} value. Similarly, write '0' time is measured as the time from when *WWL* signal is activated and storage node Q reaches 10% of V_{DD}.

3.3.5 SRAM in Standby Mode

Standby mode is the state in which there no read or write operation takes place in the SRAM cell. Inevitably, the memory cell remains in a static or hold state most of time. Due to this, the leakage power is considered as the major factor of total SRAM power consumption in subthreshold operations. In the PFC10T SRAM cell, the control signals *CS1* and *CS2* help reduce the leakage current by disconnecting the path of one of the inverters of the latch. As a result, in the hold '1' state, the leakage path from V_{DD} to *GND* is disconnected by turning M5 OFF through control signal *CS1* (refer to Figure 3.1f). Similarly, in a hold '0' state, the leakage path is disconnected by turning M6 OFF through *CS2*. This, in turn, reduces the leakage power occurred in the subthreshold SRAM cells.

3.3.6 Design Metrices of PFC10T SRAM

The RD PFC10T SRAM cell is implemented using UMC 65 nm CMOS technology using 10,000 Monte Carlo (MC) simulations in 6σ process variations in iso-area SRAM architectures. Post-layout simulation is carried out to determine various constraints like leakage current, power, read-write delay and power, read-write energy, RSNM, HSNM, WSNM, and WTP. The comparison of various constraint is performed for existing SRAM cells such as 6T, 7T [10], a differential data-aware power-supplied (D2AP) 8T [11], low-power (LP) 8T [12], and ST 10T [9]. Furthermore, all the listed constraints are observed at different temperature values ranging from 25°C to 125°C and at different process corners (PCs), namely, fast-fast (FF), slow-slow (SS), typical-typical (TT), slow-fast (SF), and fast-slow (FS) using Monte Carlo simulation. Furthermore, the observed parameters are compared with existing SRAM cells as well as with a 6-input LUT and a 2-kb macroblock. Subsequently, various parameters are described in subsections.

3.3.6.1 Cell Current

The cell current is measured at the *RBL* when *RWL* is HIGH, and a drop of minimum sensing voltage is sensed by the SA across the *RBL* [20]. Figure 3.6 shows the

SRAM Bitcells over Conventional Memories

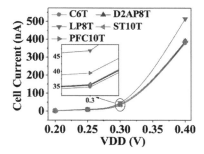

FIGURE 3.6 Comparison of cell current or read current of various bitcells at different supply voltages.

comparison of the cell current with other SRAM cells. The figure represents the cell current of a PFC10T SRAM, showing similar results at different supply voltages. The cell current is the important parameter to determine the read performance and read failure of an SRAM cell. With more cell current, the read access time is better, and more numbers of the SRAM cell can be shared with a single column of SRAM array.

3.3.6.2 Leakage Current and Power

Leakage current is premeditated as the current drawn through the metal-oxide semiconductor (MOS) devices from V_{DD} to GND while the SRAM cell is at static or hold condition. The static power or leakage power is the amount of power dissipated in the hold state. Figure 3.7 shows that a PFC10T SRAM provides a considerable amount of reduction in leakage power compared to other existing SRAM cells. This colossal amount of leakage power reduction is achieved by feedback cutting of the latch in a hold state. The control signals *CS1* and *CS2* disconnect the path between V_{DD} and *GND* and eventually reduce the flow of leakage current. Figure 3.8 shows the outcome of temperature variations on leakage power of the featured cell compared to other cells. It is observed that the leakage power of other SRAM cells increases with temperature, whereas the featured 10T SRAM cell shows better resilience against temperature variations.

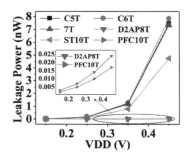

FIGURE 3.7 Comparison of leakage power of various bitcells at different supply voltages.

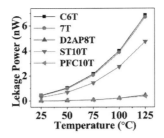

FIGURE 3.8 Comparison of leakage power of various bitcells at different temperature conditions.

3.3.6.3 Hold SNM

Hold static noise margin (HSNM) is measured when the SRAM cell is in the hold state. The HSNM is measured by varying the voltage at node QB linearly and plotting its effect on node Q and vice-versa at the worst-case PC, that is, FS. The PFC10T bitcell provides non-symmetricity at the hold state due to the feedback cutting of the latch using control signals *CS1* and *CS2*. Due to the asymmetrical architecture of SRAM at the hold state, the HSNM curve does not form a butterfly curve as shown in Figure 3.9a. It shows that the storage node Q does not flip the state with respect to the noise gained by the complementary storage node QB. However, in conventional 6T and RD8T SRAM, the storage node Q flips after the noise at storage node QB reaches $V_{DD}/2$. For holding logic 1 at storage node Q, the control signals *CS1* and *CS2* are kept at logic 0 and 1, respectively. Figure 3.9a shows the plot of HSNM at 0.3V V_{DD}, and Figure 3.9b shows the plot of the comparison of HSNM values of various SRAMs at different supply voltages. It shows that PFC10T SRAM shows significant improvement in HSNM as compared to the existing SRAM cells.

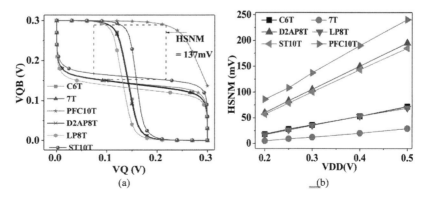

FIGURE 3.9 Comparisons of different bitcells for (a) HSNM at 0.3V V_{DD} and (b) HSNM at different supply voltages.

SRAM Bitcells over Conventional Memories

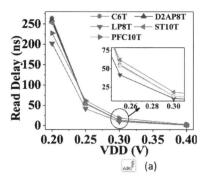
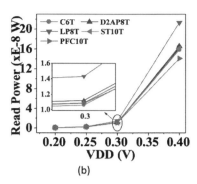

FIGURE 3.10 Comparison of different bitcells' (a) read delay and (b) read power at various subthreshold voltages.

3.3.6.4 Read Delay and Power Analysis

Read delay is measured when *RWL* is activated and *RBL* discharges and reaches the minimum sensing offset voltage required by the SA [19]. The power measured for the read operation cycle is defined as a read power. Figure 3.10a shows the plot of read delay, and it can be noticed from the figure that the PFC10T SRAM achieves similar read access compared to other existing SRAMs. Figure 3.10b shows the comparison of read power of different SRAMs. It can be observed from the figure that the featured 10T bitcell consumes less read power compared to other existing cells.

3.3.6.5 RSNM and Dynamic Read Margin

RSNM is measured by applying a DC noise voltage source at one of the storage nodes, Q or QB, and investigating the effect on other storage nodes. The RSNM is examined in the read operation when *RWL* is HIGH and *WWL* is LOW. The control signals *CS1* and *CS2* handle the read operation and improve the static noise margin. The PFC10T SRAM shows asymmetrical behavior for read operation due to the feedback cutting of the latch using control signals *CS1* and *CS2*. Due to the asymmetrical architecture of SRAM, the RSNM curve does not show the butterfly curve as shown in Figure 3.11a. It shows that storage node Q does not flip the state even with the noise gained by the complementary storage node QB. While in conventional 6T and RD8T SRAM cells, storage node Q flips when the noise becomes more than $V_{DD}/2$. The control signal disconnects one of the inverter paths, which would further help neglect the consequences of static noise and, as a result, improves the RSNM. The RSNM of the bitcell is compared with other existing cells at $V_{DD} = 0.3V$ in Figure 3.11a. The RSNM of the PFC10T SRAM is 137mV at 0.3V, and a comparison between the RSNM values of various SRAM cells is shown in Figure 3.11b.

3.3.6.6 Write Delay and Power Analysis

The write '1' access time is measured as the time when *WWL* is triggered ON and storage node Q reaches 90% of the V_{DD} value. Additionally, the write '0' access time

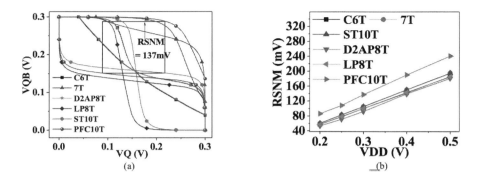

FIGURE 3.11 Comparison of RSNM of various bitcells (a) at 0.3V V_{DD} and (b) at different supply voltages.

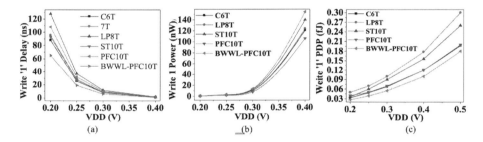

FIGURE 3.12 Comparison of various SRAMs' write '1' (a) delay, (b) power, and (c) energy at different supply voltages.

is defined as the time when the *WWL* signal is activated and storage node Q reaches 10% of the V_{DD} value. The write dynamic power is measured as the product of the average current flow and the source voltage at the write access time. Figure 3.12a shows the observed write '1' delay at various supply voltages. The delay of the PFC10T SRAM is compared with existing cells, which shows that the 10T cell provides faster write access compared to LP8T SRAM at different supply voltages, as shown in Figure 3.12a. Furthermore, the write delay is also observed for boosted write wordline (*BWWL*) at $1.2 \times V_{DD}$. It can be seen that using *BWWL*, the cell writes the bit faster compared to existing SRAMs. Similarly, Figure 3.12b demonstrates write '1' dynamic power values at various supply voltages. It is observed that the cell consumes the least write dynamic power compared to existing cells. Furthermore, Figure 3.12c shows the PDP or energy for the write '1' operation. It indicates that the PFC10T bitcell has a lower energy consumption compared to other existing cells. In addition, the *BWWL* PFC10T SRAM shows the most energy-efficient results. Thus, the low write energy makes the PFC10T bitcell energy-efficient at the subthreshold voltages. This, in turn, makes the PFC10T SRAM as the best choice for memory design for FPGA.

3.3.6.7 WSNM

The WSNM is measured at the time of write operation by initiating a linear DC noise at one of the storage nodes and observing its effect on the opposite storage node. Figure 3.13a shows the WSNM curve of 10T SRAM compared with existing SRAM architectures. The WSNM value of 10T SRAM is measured as 216mV at $V_{DD} = 0.3$V, which shows a higher value than other existing SRAMs. In addition, Figure 3.13b shows that the PFC10T SRAM cell offers significant improvement in WSNM values as compared to other existing SRAMs at different supply voltages.

3.3.6.8 WTP

WTP is measured at the time of write operation while *WWL* is HIGH and *RWL* is LOW. It is observed by adding a linearly variable source at *WBL* and observing its effect on *WBLB*. The WTP is derived obtained as the difference between V_{DD} and the crossover point of storage node Q and QB. The WTP for the PFC10T bitcell comes out to be 166.5mV at 0.3V power supply. Figures 3.14a and 3.14b show the plot of WTP and WTP% with respect to V_{DD} at different power supplies, respectively. It shows improved results in terms of WTP or WTP% with respect to V_{DD} compared to other existing SRAM cells.

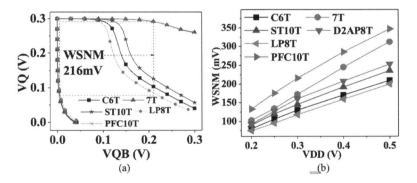

FIGURE 3.13 Comparisons of (a) WSNM at 0.3V power supply and (b) WSNM at different supply voltages.

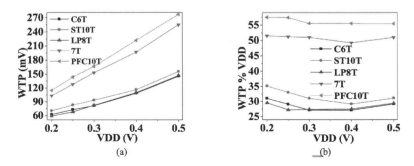

FIGURE 3.14 Comparisons of (a) WTP and (b) WTP% with respect to V_{DD}.

3.3.7 Discussion on Post-Layout Simulation Results

Table 3.3 shows the summary of post-layout simulation results of the PFC10T SRAM cell at 0.3V power supply. The parameters are observed at 27°C room temperature, where all transistors are taken as a low-V_{th} transistor. The bitcell WSNM is improved by 1.66×, 1.44×, 1.83×, and 1.49×; RSNM by 3.8×, 1.54×, 1.37×, and 1.3×; WTP by 2×, 1.87×, 2×, and 1.78× compared to 6T, D2AP8T, LP8T, and ST10T SRAM cells, respectively. The PDP for a write '1' operation is reduced by 173%, 11.11%, 40%, and 22% as compared to 6T, D2AP8T, LP8T, and ST10T SRAM cells, respectively. The leakage power is also reduced to 0.07×, 0.7×, and 0.0125× compared to 6T, D2AP8T, and ST10T SRAM cells, respectively, at 0.3V V_{DD}. The presented 10T bitcell also shows a better I_{on}/I_{off} ratio to 71.39×, 34.67×, 1.49×, and 46.47× compared to 6T, 7T, D2AP8T, and ST10T SRAM cells, respectively. Furthermore, Table 3.4 illustrates the values of SNMs at different PCs. It can be observed from the table that all cells have the worst HSNM and RSNM values at the SF PC. In addition, the WSNM has a worst-case value at the FS PC. The benefit of the featured bitcell is its susceptibility toward various process corners.

3.3.8 Application of PFC10T SRAM in FPGAs

The PFC10T SRAM cell is employed to implement a 6-input LUT and a 2-kb macroblock. This section demonstrates the simulation results of the LUT and macroblock. Furthermore, the results are compared with existing SRAM cell–based LUTs and macros. The LUT and macro show faster access time, low leakage power consumption, and low PDP or energy.

3.3.8.1 6-Input LUT

The 6-input LUT is implemented using a UMC 65nm CMOS technology node. The controller logic is used before each SRAM cell to improve the SNM and leakage power. The six-input LUT consists of 64 SRAM cells and a 64 × 1 multiplexer as shown in Figure 3.15. The basic principle of FPGAs is considered a reconfigurable logic. While programming a digital system, the favorable outputs corresponding to

TABLE 3.3
Summary of Various Constraints Considered for the Comparisons with the PFC10T SRAM

SRAM	HSNM (mV)	WSNM (mV)	RSNM (mV)	RDNM (mV)	WTP (mV)	Leakage Power	I_{on}/I_{off}	Write '1' PDP (fJ)	Write '0' PDP (fJ)
6T	100	130	36	263	81.3	1.28nW	22.1	0.246	0.23
D2APT	92	150	35	262	89	14.2pW	1060	0.1	0.1
LP8T	100	118	100	273	82	7.65pW	3130	0.126	0.1
ST10T	105	145	105	262	93.2	0.8nW	34	0.11	0.1
PFC10T	137	216	137	300	166.5	10pW	1580	0.09	0.07

SRAM Bitcells over Conventional Memories

TABLE 3.4
WSNM, RSNM, and HSNM Mean (μ) at Different PCs Using Monte Carlo Iterations

SRAM	Process SNM (mV)	FF	SS	TT	SF	FS
6T	HSNM	104	100	36	86.5	75
	RSNM	31	37.7	100	63	Unstable
	WSNM	125	118	130	35	138
D2AP8T [11]	HSNM	100	98.6	92	87.3	71
	RSNM	30	38	35	62.3	10
	WSNM	160	150	160	190	154
LP8T [12]	HSNM	100	97.6	100	93	67.7
	RSNM	100	97.6	100	93	67.7
	WSNM	122	113.6	118	30	140
ST10T [9]	HSNM	106.5	105.5	105	76	87.7
	RSNM	106.5	105.5	105	76	87.7
	WSNM	150	147.4	145	30.8	163.5
PFC10T	HSNM	139.7	132.7	137	100	133
	RSNM	139.7	132.7	137	100	133
	WSNM	218	195	216	203	260

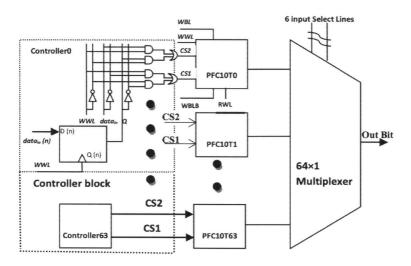

FIGURE 3.15 The architecture of PFC10T SRAM–based 6-input LUT.

the inputs are stored in the SRAM cells of LUT. After programming the FPGA, we can reconfigure the same logic based on the stored output values in an SRAM cell and the input select lines. The simulation results of LUT are observed at different subthreshold regions.

Figure 3.16a demonstrates the comparison of leakage power at different supply voltages. The leakage in the 6-input LUT is reduced significantly as compared to the existing LUTs. Furthermore, Figure 3.16b explains the plot of read '1' power consumption at different V_{DD} values. It can be noticed that the PFC10T SRAM cell–based LUT attained a minimum read power dissipation. The read '1' power is observed as the power utilization when the six-input select lines are activated and transmit the stored information in the SRAM cell to the output of LUT. The read power is measured as the product of read current and V_{DD} when 90% of information is read at the output of LUT. Figures 3.17a and 3.17b exhibit the write '1' delay and write '1' power, respectively. It shows a similar

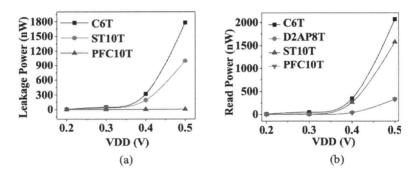

FIGURE 3.16 Comparison of (a) leakage power and (b) read power at various supply voltages of LUTs.

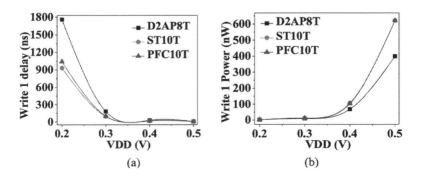

FIGURE 3.17 Comparison of (a) write '1' delay and (b) write '1' power at various supply voltages of LUTs.

SRAM Bitcells over Conventional Memories

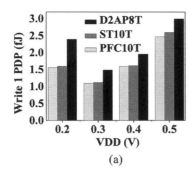
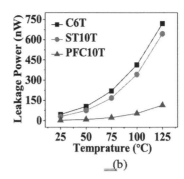

FIGURE 3.18 Comparison of (a) write '1' PDP and (b) leakage power at various temperatures of LUTs.

TABLE 3.5
Summary of Various Constraints and Comparisons with a PFC10T-Based LUT

SRAM parameter	ST10T [9]	D2AP8T [11]	PFC10T
Leakage power (nW)	32.2	6.5	2.64
Write '1' delay (ns)	93.18	183.2	101.6
Write '1' power (nW)	12.07	8.13	11.35
Write '1' PDP (fJ)	1.12	1.49	1.1

characteristic as compared to D2AP8T, ST10T, and PFC10T cells, which illustrates that the LUT has achieved better leakage power and cell stability with a similar outcome in read-write delay and power. Furthermore, Figure 3.18a shows write '1' PDP at different V_{DD} values, and it is clear that the PFC10T bitcell LUT provides minimum energy consumption as compared to existing cells. Moreover, Table 3.5 illustrates the summary of results at 0.3V V_{DD} in the TT PC. The results are from 10,000 Monte Carlo iterations, considering the effect of different process variations.

The presented LUT shows a reduction in leakage power by 12.2× and 2.46× compared to ST10T and D2AP8T SRAM cells, respectively. It also shows a reduction in write '1' PDP by 2% and 35.45% compared to ST10T and D2AP8T SRAM cells, respectively. The leakage power with respect to the temperature is shown in Figure 3.18b, where the LUT shows least number variations with respect to the temperature. The leakage power of LUT is also examined at various PCs in Figure 3.19. The figure demonstrates the effect of process variability has the least impact on LUT. The FF comes out to be the worst-case PC, and still, it exhibits significantly better leakage power as compared to existing cells.

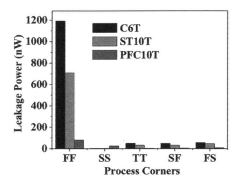

FIGURE 3.19 Leakage power of six-input LUT at different PCs at 0.3V V_{DD}.

3.4 SUBTHRESHOLD PROCESS TOLERANT 10T SRAM FOR IOT APPLICATIONS

In this section, an ultra-low-power (ULP) 10T SRAM cell is presented for the application of IoT low-end edge devices. The low-end IoT devices have exceptionally low power requirements. Therefore, the featured bitcell is designed in the subthreshold voltage power supply. However, due to voltage scaling and working at the subthreshold voltages, a very high leakage power arises. The issue of leakage power in the subthreshold SRAM also occurred with variations in various process parameters and temperature values. In addition, the unpredictability in device performance, instability due to noise generated from high bitline static voltage and thermal noise generated from temperature variations caused SRAM device failure at subthreshold operations. Therefore, to overcome the aforementioned issues, a differential process tolerant 10T (PT10T) SRAM cell is shown here.

3.4.1 PT10T SRAM Bitcell

The advancement in technology and the reduction in integrated circuit (IC) development costs due to smaller die areas resulted in considerable improvement in digital electronic devices. If this development continues, it will result in a world where all digital electronic devices are linked to the internet, known as the IoT. Furthermore, a million of zettabytes of memory is required to handle the large-scale computations required by IoT devices. Thus, memory plays a vital role in future energy-efficient computing systems, and therefore, the subthreshold, low-leakage, and high-stability SRAM is presented in this section. The featured cell shown in Figure 3.20 has two-write access n-MOS transistors, MN1 and MN3, and two read-access transistors, MN2 and MN4. Bitline (*BL*) contains single-bit information to write, and bitline bar (*BLB*) contains the complementary of that. In addition, n-MOS transistors MN2 and MN4 are connected with virtual ground n-MOS transistor MN_GND. *BL* and *BLB* are charged to V_{DD} before the read operation is performed. MP1 and MP2 and MN5 to MN8 form a latch, where MP1 and MP2

SRAM Bitcells over Conventional Memories

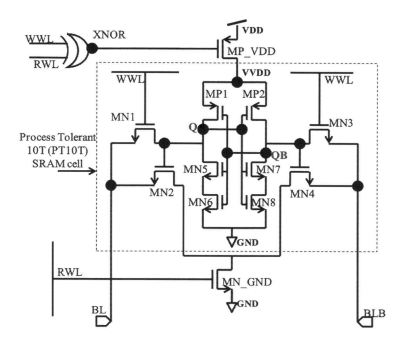

FIGURE 3.20 Circuit diagram of a PT10T SRAM bitcell.

are the pull-up transistors linked to virtual V_{DD} (VV_{DD}). In addition, MN5–MN6 and MN7–MN8 form a stack, which eventually enhances the read and write noise margin. Furthermore, two transistors, namely, MP_V_{DD} and MN_GND, are shared with each row of 8-kb SRAM, as shown in Figure 3.21, where MP_V_{DD} is used as a power-gating p-MOS transistor, which disconnects the path between V_{DD} and *GND* at the holding state to improve the leakage power. Furthermore, Table 3.6 explains the working principle and operation of 10T SRAM. Also, the operations of bitcell at various modes namely, read, write, and standby are explained in Section 3.4.2. Furthermore, the comparison between the layout areas is demonstrated in Table 3.7. The area overhead of PT10T SRAM is 135% of conventional 6T SRAM, which eventually reduces the overall memory density. However, the improvements in other key parameters such as reduction of leakage power and increments in write and read stability (which are discussed in subsequent sections) can lead to using PT10T SRAM for IoT applications, where power, reliability, and accuracy are the key concerns.

3.4.2 Operations of PT10T SRAM

This section demonstrates the working principle of 10T SRAM in subthreshold operations. Furthermore, the operation has been performed in 8-kb SRAM array. The read and write operations are carried out from a 64-bit SRAM column.

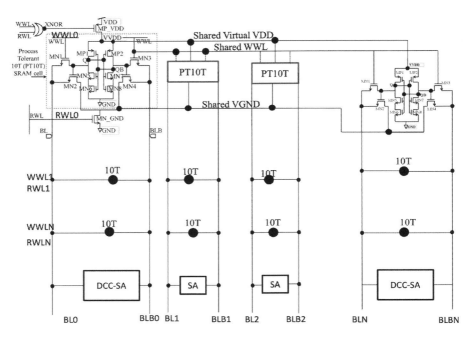

FIGURE 3.21 Simplified array architecture of 10T SRAM, where the read is performed using a high-speed differential current compensation sense amplifier (DCC-SA).

TABLE 3.6
Logic Truth Table for Various Operations in PT10T SRAM

Operation	WWL	RWL	BL	BLB
Write 1	1	0	1	0
Write 0	1	0	0	1
Read 1	0	1	Discharging	1
Read 0	0	1	1	Discharging
Hold	0	0	1	1

TABLE 3.7
Comparison of Cell Layout Area in UMC 65nm CMOS Technology

SRAM	Number of LVT transistors used	Layout area	Normalized layout area
6T	6 (2 LVT PMOS, 4 LVT NMOS)	2.33µm^2	1×
RD8T [8]	8 (2 LVT PMOS, 6 LVT NMOS)	2.8µm^2	1.2×
ST10T [9]	10 (2 LVT PMOS, 8 LVT NMOS)	3.68µm^2	1.58×
LP10T [22]	10 (2 LVT PMOS, 8 LVT NMOS)	4.0µm^2	1.71×
PT10T	10 (2 LVT PMOS, 8 LVT NMOS)	3.15 µm^2	1.35×

SRAM Bitcells over Conventional Memories

3.4.2.1 Read Operation

The read operation is obtained through ultra-fast differential current compensation SA (DCC-SA) [23]. The read operation is performed by keeping *RWL* to HIGH and *WWL* to LOW and XNOR_OUT kept at a LOW value as shown in Figure 3.22a. For a read '1' operation, logic 1 is stored at storage node Q and *RWL* is kept at HIGH, which eventually turns ON MN_GND and MN2. This forms a discharging path across BL–MN2–MN_GND and a voltage difference, $\Delta V_{BL} = \{(V_{DD} - (V_{DD} \cdot I_{read \cdot BL} \times R_{MN2\_GND})\}$ appears between *BLs* which is sensed by the full swing inverter sense amplifier, where I_{read} is the cell current and $R_{MN2-GND}$ is the resistance through MN2 and MN_GND. The read access time is measured as the time the *RWL* signal is activated until the *BLB* is discharged to the minimum required potential needed by SA to read. The sensing voltage ΔV_{BL} required for DCC-SA is 80mV [23]. The read access time and power are measured across a 64-bit column of the SRAM cell. The parasitic capacitance of an SRAM column is measured after RC extraction in 65nm CMOS technology.

3.4.2.2 Write Operation

For a write operation, *WWL* must be HIGH and *RWL* must be LOW. However, by making *WWL* HIGH and *RWL* LOW, *XNOR_OUT* is switched to a LOW value. The logic 1 is written to storage node Q through BL–MN1–Q as shown in Figure 3.22b. However, the write '1' access time is measured as the time when the *WWL* signal is activated and reaches its $V_{DD}/2$ value and storage node Q reaches 90% of the V_{DD} value. Similarly, the write '0' time is measured as the time when the *WWL* signal is activated and reaches its $V_{DD}/2$ and storage node Q reaches 10% of V_{DD}. The half-select issue in the write operation is also taken care of by putting a *WE* signal at each column of the SRAM array (refer to Figure 3.21).

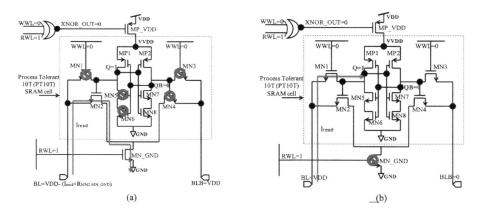

FIGURE 3.22 (a) State of read '1' operation of a PT10T SRAM cell. (b) State of write '1' operation of PT10T SRAM, where the power gated PMOS MP_V_{DD} is turned ON by the external XNOR gate.

By making the read and write paths separate using *WE* and *REB*, the sensing and writing operations are separated, which eventually helps counteract the half-select read and write issues.

3.4.2.3 Data Retention

In the PT10T SRAM, the control signal XNOR_OUT turns OFF MP_V_{DD}, which helps reduce the leakage current by disconnecting the path of the latch from V_{DD} to *GND*. The leakage power in a 6T SRAM cell introduced is due to the non-availability of virtual V_{DD} transistors such as MP_V_{DD}. The standby or leakage power dissipation is one of the major problems with embedded cache in sub-nanometer technology nodes. Leakage power is the major contributor to the total power consumption of SRAM, as most of the cells remain in the idle state. The leakage current has mainly three components namely, gate leakage, junction leakage, and subthreshold leakage through different transistors [24]. If two devices are connected in series, then they form a stack, and if one terminal of a stack is connected to V_{DD} and the other end is connected to the ground, then the intermediate node rises to a voltage that is higher than the ground potential.

In the featured SRAM, at a hold mode, as shown in Figure 3.23, the master transistor (MP_V_{DD}) is turned OFF as output of the XNOR-gate is HIGH (due to both *WWL* and *RWL* are LOW). Therefore, the cross-coupled inverters are decoupled from the V_{DD}, and a stack is formed using MP1, MP2, and MP_V_{DD}.

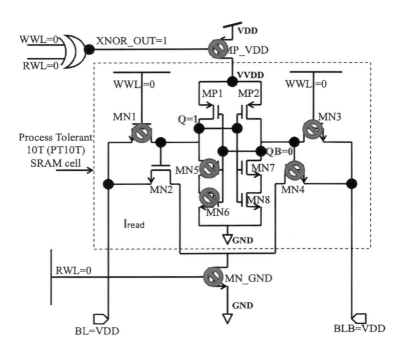

FIGURE 3.23 Schematic of PT10T SRAM cell at hold mode.

Because of this effect, the intermediate node VV_{DD} rises to some positive voltage [25]. This positive voltage reduces leakage and holds power during standby mode. However, in 6T and RD8T SRAM cells, there is a potential difference in the V_{DD} voltage between the V_{DD} and *GND* terminals. However, due to the power gating of V_{DD}, the virtual node VV_{DD} now is set to a positive voltage lower than V_{DD}. This eventually reduces the HSNM but improves the leakage power required for IoT edge devices.

3.4.3 PT10T Write and Read Analysis

The write '1' access time is measured as the time when the *WWL* signal is triggered and storage node Q reaches 90% of the V_{DD}. Similarly, write '0' access time is defined as the time when the *WWL* signal is activated and storage node Q reaches 10% of the V_{DD}. The write power is measured as the product of average current dissipation and the source voltage till write is achieved. Figure 3.24a shows the comparison of write '1' delay of various bitcells observed at different supply voltages in the worst-case (SS) PC. The delay of LVT PT10T SRAM is compared with existing LVT 6T, RD8T [8], LP9T [21], and LP10T [22], which shows that the PT10T SRAM has a comparable write access time with respect to other bitcells. Similarly, Figure 3.24b show comparisons of write '1' power of various bitcells at different supply voltages in FF PC. The results obtained show that the PT10T has less write '1' power by 8.5%, 12.6%, 19.35%, and 19.3% compared to 6T, RD8T [8], LP10T [22], and LP9T [21] SRAM, respectively, at $V_{DD} = 0.3$V.

Furthermore, the read access time is measured when *RWL* is activated. From the DCC-SA architecture presented by Reniwal et al. [23], it is observed that a differential sensing voltage of 80mV is required to read data from SRAM. Figure 3.25a shows the simulation states of SRAM at read time. The read access time of the PT10T SRAM is compared with the existing SRAM cells as shown in Figure 3.25b. It is observed that the read access of LVT-based PT10T SRAM is 1.6×, 1.57× faster than 6T and ST10T cells [9], respectively, in SS PC at $V_{DD} = 0.3$V.

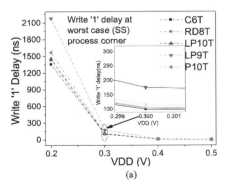
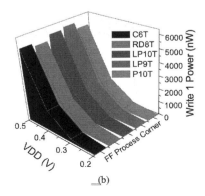

FIGURE 3.24 Comparison of (a) write '1' delay and (b) write '1' power at FF PC.

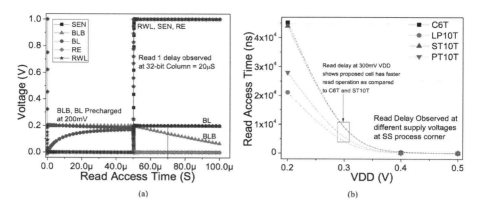

FIGURE 3.25 (a) State of input variables in the read operation. (b) Read access time at the SS corner at different voltages.

3.4.4 PT10T Leakage Power

Leakage current is measured as the current drawn from V_{DD} to GND while the SRAM cell is in a static or hold condition. The static power or leakage power is the amount of power dissipated at the hold state. Figure 3.26a shows the leakage power variations with respect to the supply voltages for already existing and featured PT10T SRAM. However, after determining the leakage power of 10T SRAM comes out to be 10.5pW, which is 0.012×, 0.013×, and 0.93× the 6T, RD8T, and LP9T SRAM, respectively, at 0.3V V_{DD}. In addition, Figure 3.26b shows the occurrences of leakage power variations of the PT10T bitcell at different supply voltages, which reduces the peak value of leakage power to sub-100pW at different supply voltages.

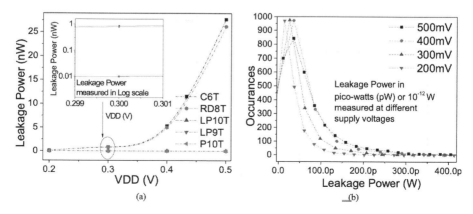

FIGURE 3.26 (a) Comparison of leakage power at different supply voltages and (b) the distribution of leakage power of PT10T SRAM at various supply voltages.

3.4.5 PT10T RSNM

The RSNM is measured by applying a DC noise voltage source at one of the storage nodes, Q or QB, and investigating its effect on other storage nodes. The RSNM is examined in the read operation when *RWL* is HIGH and *WWL* is LOW. The decoupled read path through BL–MN2–MN_GND does not affect the storage nodes of the SRAM cell, which would further help negate the consequences of static noise and, as a result, improve the RSNM. The RSNM of the PT10T SRAM cell comes out to be 107mV, which shows 3×, 1.48×, and 1.48× improvement compared to 6T, RD8T, and LP10T SRAM, respectively, at $V_{DD} = 0.3$V as shown in Figure 3.27.

3.4.6 PT10T WSNM

The WSNM is measured at the time the write operation by initiating a linear DC noise at one of the storage nodes and observing its effect at the other end. For the WSNM measurement, the *WWL* is kept HIGH and the *RWL* is LOW. The plots in Figures 3.28a, 3.28b, 3.28c, and 3.28d determine the WSNM of 6T, LP9T, LP10T, and PT10T SRAM cells, respectively, at the SF process corner. Moreover, from Figure 3.28a and 3.28b, it is observed that the WSNM of 6T and LP9T SRAM fails to write at 0.2V with WSNM (20mV) less than the thermal voltage (28mV).

In addition, LP10T has a WSNM of 10mV and 30mV at 0.2V and 0.3V V_{DD}, respectively. Thus, it fails to operate at 0.2V and has a near-threshold value at 0.3V (WSNM = 30mV). Consequently, it is suggested to operate LP10T SRAM above 0.3V supply voltage. Furthermore, the presented 10T SRAM has a WSNM of 21mV and 40mV at 0.2V and 0.3V V_{DD}, respectively, in the SF PC. The WSNM of PT10T shows 2×, 2×, and 1.5× higher outcomes as compared to 6T, LP9T, and LP10T SRAM cells, respectively, at 0.3V V_{DD}.

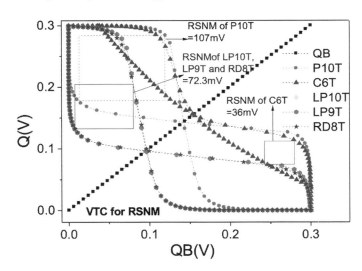

FIGURE 3.27 RSNM of different SRAMs observed at 0.3V V_{DD}.

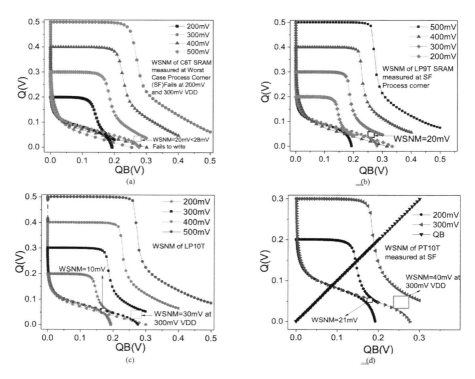

FIGURE 3.28 WSNM of (a) LP10T SRAM, (b) PT10T SRAM, (c) 6T SRAM, and (b) LP9T SRAM at different supply voltages.

3.4.7 8-kb Subthreshold SRAM for IoT Applications

Table 3.8 shows the summary of post-layout simulation results of PT10T 8-kb SRAM at 0.3V power supply. The parameters are observed at 27°C (room temperature) in the worst-case process corner. Table 3.8 shows that the leakage power of LVT based 10T SRAM is reduced by 98.76%, 98.6%, 6.7%, and 98.2% as compared to the LVT 6T, RD8T, LP9T, and ST10T SRAM cells, respectively, at 0.3V V_{DD}. The WSNM value is improved by 2×, 1.25×, 2×, 1.33×, and 1.14× as compared to 6T, RD8T, LP9T, LP10T, and ST10T SRAM, respectively, at the worst-case (SF) PC. The RSNM is enhanced by a factor of 3×, 1.48×, 1.48×, and 1.48× and 3.88% as compared to 6T, RD8T, LP9T, LP10T, and ST10T SRAM cells at the TT PC, respectively. The dynamic read margin is determined at the time of read operation and shows an improvement of 31.8% and 8.2% compared to 6T and ST10T SRAM, respectively. The write '1' energy is reduced by 6%, 2%, and 6% as compared to LP9T, LP10T, and ST10T SRAM cells, respectively, at the worst-case (SS) PC at 0.3V V_{DD}. However, the read '1' energy is reduced to 4.3%, 12%, 6.4%, 4%, and 1% as compared to 6T, RD8T, LP9T, LP10T, and ST10T SRAM cells, respectively, at the worst-case (SS) PC at 0.3V V_{DD}. By comparison, the footprint area of the bitcell has 35% and 12.5%

TABLE 3.8
Summary of Mean (μ) Values of Simulation Results of PT10T 8kb SRAM Compared with Other Existing SRAM Architectures Using 6-Sigma Process Variance at 0.3V V_{DD}

Merits SRAM	WSNM (mV)	RSNM (mV)	Leakage power	Write '1' energy (pJ)	Read energy (fJ)
6T	20	36	6.8μW	0.16	23
RD8T [8]	32	72.3	6.3μW	0.16	25
LP9T [21]	20	72.3	90nW	0.17	23.5
LP10T [22]	30	72.3	81nW	0.163	23
ST10T [9]	35	103	4.66 μW	0.17	22.2
PT10T	40	107	84nW	0.16	22

more overhead as compared to 6T and RD8T SRAM cells, respectively, whereas the PT10T SRAM has 17% and 27% less macroblock area compared to ST10T and LP10T cells, respectively. Therefore, the PT10T SRAM is an attractive choice for today's battery-operated IoT-enabled system-on-chip (SoC) applications, where the leakage power consumption and cell stability are primary concerns. Furthermore, Table 3.9 is a comparison table of all the fundamental parameters measured using different threshold voltage transistors (HVT, RVT, LVT). The inter-die and intra-die process/mismatch variations are considered at 0.3V power supply at room temperature. All the parameters have been measured at the worst-case PC. From the table, it is observed that the featured 10T SRAM shows the best performance in terms of write power, read-write energy, leakage power, RSNM, and WSNM. The objective of this research triggers creating memory architecture for ULP IoT applications can be achieved by introducing low leakage and high stability of SRAM.

3.5 ROBUST SUBTHRESHOLD 8T SRAM FOR IMAGE PROCESSING

The SRAM plays a vital role in image and signal processing. The system used for object tracking and image analysis requires high accuracy and low leakage power SRAMs while operating on battery power. Object tracking is an important aspect in the field of computer vision. There are three major steps in object tracking: detection of an object, tracking it from frame to frame, and analyzing the object's motion to identify its actions [26–29]. A substantial amount of work has been done in the past to improve the accuracy and speed of object tracking. From the previous research work, it is noted that previously, the focus was on improving the accuracy of object tracking in real-time scenarios, which introduce extra hardware, specifically SRAMs. The SRAM used in a tracking system consumes more power due to high-definition video/image compression and processing. Thus, here an 8T SRAM is offered to overcome the standby power issue with better cell stability, which improves the total power consumption required for object detection and tracking. Furthermore, a fast, reliable,

TABLE 3.9
Summary of Mean (μ) Values of Inter/Intra-Die Process Mismatch Threshold Voltage (V_{th}) Variations of PT10T SRAM Compared with Other Existing SRAM Using Different Threshold MOS Transistors at 0.3V V_{DD}

SRAM Merits	6T RVT	6T HVT	6T LVT	RD8T [8] RVT	RD8T [8] HVT	RD8T [8] LVT	ST10T [9] RVT	ST10T [9] HVT	ST10T [9] LVT	LP10T [22] RVT	LP10T [22] HVT	LP10T [22] LVT	PT10T RVT	PT10T HVT	PT10T LVT
Write '1' delay (ns) SS	683	11840	99.43	687	12600	104	786	11500	110	770	13300	107	1300	13500	116.3
Write '1' power (pW) SS	124	12.56	1.6	130	12.34	1.6	122	14	1.5	146	13	1.52	70 best	10 best	1.5 best
Write '1' energy (pJ) SS	0.85	1.48	0.16	0.89	1.55	0.16	0.96	1.61	0.17	1.12 worst	1.73 worst	0.163 worst	0.78 best	1.35 best	0.16 same
Read delay SS (μs)	138	4000	2.3	174	5000	3	130	4000	2.3	50	1620	1	53	1650	1
Read power SS (pW)	150	5.23	10000	144	5.12	8330	151	5.22	9660	405 worst	13 worst	23000 worst	390	12.7	22000
Read energy SS (fJ)	20.7	20.9	23	25	25.6	25	19.6	21	22.2	20.2	21	23	20.6	20.9 best	22 best
Leakage power (nW)	752	72	6800	640	64	6288	552	46.4	4660	64	8	80	80	20	84
RSNM (mV) FS	10	21.5	36	100	80	72	97	78	103	87	68	72	103 best	83 best	107 best
WSNM (mV) SF	20	Fails	20	28	Fails	30	33	Fails	35	30	33	30	40 best	30	40 best

SRAM Bitcells over Conventional Memories

less memory-demanding object detection and tracking algorithm is presented here. The goal is achieved in four steps, namely, segmentation and thresholding, object detection, quadtree method to minimize pixelation, and tracking. Furthermore, the performance of the 8T SRAM is examined by computing the average memory size required for the object tracking device.

3.5.1 PFC Subthreshold 8T SRAM

Object tracking requires a ULP memory block to store the information of present and reference macroblock in the form of pixels. Each gray pixel has 8-bit information. Therefore, an 8-bit array of memory is required for storing one pixel. The memory array is designed using SRAM cells due to its fast write and read access time. However, there are some limitations of the subthreshold SRAM-based memory architecture, specifically, high leakage power, low static noise margins, disparity in stability, and susceptibility due to different PVT conditions. Consequently, a read-decoupled PFC 8T (PFC8T) SRAM cell is presented here to resolve the ambiguous behavior of the subthreshold SRAM architectures at different PVT conditions in UMC 65nm standard CMOS technology. The schematic of an 8T cell and its layout are shown in Figure 3.29a and Figure 3.29b, respectively. All the MOS transistors in the presented and existing cells are taken as a low-voltage-threshold (LVT) transistor. However, the featured 8T cell attains 1.4× more cell layout area as compared to the 6T SRAM cell; the improvement in leakage power and cell stability at different PVT conditions makes it an ideal choice for portable memories. The PFC8T bitcell has two write access NMOS transistors, MN1 and MN2. An input bit is written to the SRAM cell through these transistors. The NMOS transistors MN4 and MN6 are used as a read decoupled logic to read information from the cell. The MP1, MP2, MN3, MN4, and MN5 transistors form a latch, where MP1 and MP2 are the

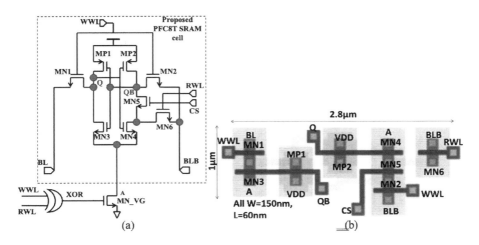

FIGURE 3.29 (a) The PFC8T SRAM bitcell and (b) the layout of PFC8T SRAM cell at 65nm CMOS technology.

TABLE 3.10
Truth Table for Various Operations in PFC8T SRAM

Operation	WWL	RWL	BL	BLB	CS	XOR I/P
Write 1	1	0	1	0	1	1
Write 0	1	0	0	1	1	1
Read 1	0	1	0	0	0	1
Read 0	0	1	0	0	0	1
Hold 1	0	0	0	0	1	0
Hold 0	0	0	0	0	1	0

pull-up PMOS transistors linked to V_{DD}. In addition, MN3 and MN4 are the pull-down transistors connected with MN_VG, where MN_VG is an NMOS transistor connected to the ground terminal, which is eventually used as a stacking transistor to improve the leakage power in hold operation. The MN_VG is controlled by an external *XOR* gate, which is shared among each row of the SRAM. Moreover, MN5 is the feedback-cutting transistors that is used to disconnect the path between V_{DD} to *GND*. This transistor is activated by control signal *CS*, which helps improve read static noise margins and leakage power of the cell. The operations at different states are presented in Table 3.10.

3.5.2 SRAM LAYOUT AND MACROBLOCK

A 2 × 2 array layout, including input/output ports and MOS transistors, is shown in Figure 3.30. The layout shows the connection between various input and output ports connected using different metals. In 8T bitcell architecture, three metal layers are used, namely, M1, M2, and M3. The design of an array is made as compact as possible. These metal layers are connected with each other and poly through 'via', such as Poly–M1, M1–M2, and M2–M3. There are separate *WWL* and *RWL* for each row of an array to control write and read operations, respectively. The control signal (*CS*) is employed to separate the read path from the latch. The *BL* and *BLB* are also shared among the column of an array. However, the bitline capacitance associated with the read and write operations for an SRAM layout depends on the number of bits linked with each column. In our case, the bitline in the read condition shares two transistors, MN2 and MN6. Due to two pass transistors coupled to the read path, the bitline capacitance is increased. Although the capacitance is higher than that of conventional 6T SRAM and RD8T SRAM, the leakage power and static noise margin are improved with the PFC8T bitcell. In addition, the cell layout area of RD8T is similar to the 8T SRAM. The architecture of an 8T SRAM macro is shown in Figure 3.31. The macro consists of 32 rows and 64 columns. The *XOR* gate is shared with each row of an SRAM array to offer an input to the MN_VG transistor. The stack transistor MN_VG is linked with each row, which eventually reduces the leakage power.

SRAM Bitcells over Conventional Memories

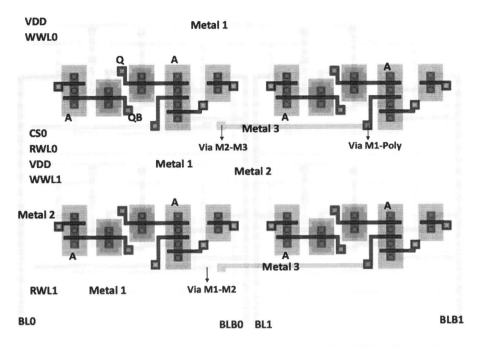

FIGURE 3.30 PFC8T SRAM cell 2 × 2 array layout. The *RWL*, *WWL*, *CS*, and V_{DD} are shared at each row, while *BL* and *BLB* are shared at each column.

3.5.3 OPERATIONS OF PFC8T SRAM

Various operations like read, write, and hold are explained in this section.

3.5.3.1 Read Operation

Read operation is performed by keeping read wordline (*RWL*) HIGH, write wordline (*WWL*), and the control signal (*CS*) LOW, as shown in Figure 3.32a. Subsequently, for read '1' (Q=1), logic 1 is stored at storage node Q, and *RWL* is kept HIGH, which eventually turns ON MN4 and MN6. This forms a discharging path across BLB–MN6–MN4–MN_VG and a voltage difference, $\Delta V_{BLB} = \{V_{DD} - [V_{DD} - I_{read} \times R_{MN\_VG-MN4-MN6}]\}$ appears between *BL* and *BLB*, which is sensed by the full-swing inverter SA, where I_{read} is the cell current and $R_{MN\_VG-MN4-MN6}$ is the resistance through MN_VG, MN4, and MN6.

3.5.3.2 Write Operation

For a write operation, *WWL* is kept HIGH and the *RWL* is kept LOW. To write '1', the control signal (*CS*) must be HIGH. The logic 1 is written to storage node Q through BL–MN1–Q. The write '1' time is measured as the time when the *WWL* signal is HIGH, and storage node Q reaches 90% of the V_{DD}. Similarly, write '0' time is measured as the time when the *WWL* signal is HIGH and node Q reaches 10% of the V_{DD}. The write '1' operation is shown in Figure 3.32b.

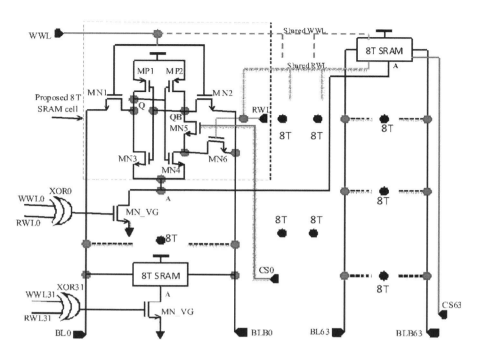

FIGURE 3.31 The architecture of 8T SRAM cell–based 32 × 64-bit macroblock.

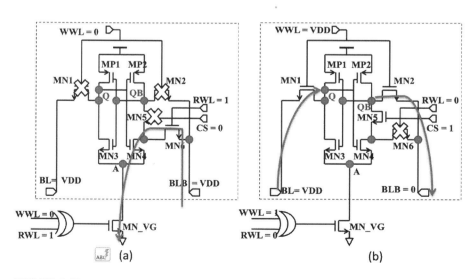

FIGURE 3.32 (a) Read operation schematic shows discharging of *BLB* potential from V_{DD} to V_{DD}-V_T, where V_T is the minimum voltage required by SA to sense and read from the cell, and (b) a write '1' operation schematic shows charging and discharging of storage nodes Q and QB.

SRAM Bitcells over Conventional Memories 109

3.5.3.3 Standby Power Estimation

Generally, the memory cell remains in a static or hold state, therefore, there would be an increase in leakage power in a SRAM cell at different PVT values. In a PFC8T SRAM cell, by keeping *WWL* and *RWL* LOW, MN_VG gets turned OFF, which helps reduce the leakage current by disconnecting the path of the latch from the V_{DD} to the *GND*. The leakage power in a 6T SRAM cell is introduced due to the non-availability of the virtual ground transistor such as MN_VG. In the 8T SRAM (refer to Figure 3.31), the MN3/MN4 and MN_VG transistors form a stack, where one node of the stack is linked to the V_{DD} and the other end is connected to the ground. In hold mode, the tail transistor (MN_VG) gets turned OFF as an output of the *XOR*-gate is low (due to both *WWL* and *RWL* being LOW). Therefore, the cross-coupled inverters are decoupled from the ground, and a stack is formed between MN3/MN4 and MN_VG. Because of this effect, the intermediate node A rises to a positive voltage. This positive voltage (V_A) reduces the leakage power by reducing the potential difference between V_{DD} and *GND* to read V_{DD}-V_A.

3.5.3.4 RSNM

A key figure of merit for an SRAM cell is its RSNM. It can be extracted by plotting the largest possible square in the two voltage transfer curves (VTC) of the involved CMOS inverters [30]. The RSNM is defined as the length of the side of the square, given in volts. When external DC noise becomes more than the RSNM value, the state of the SRAM cell will change, and the data will get corrupted. In a read '0' operation, nodes Q and QB are at logic 0 and logic 1 state, respectively. In a conventional 6T SRAM, DC noise is added at Q or QB, which flips the state of the opposite storage node and causes a reduction in RSNM value. Moreover, in the PFC8T SRAM read operation, *CS* is LOW, which eventually turns OFF MN5. This subsequently disconnects the path from QB to *GND* and improves the RSNM value. Consequently, QB will remain at logic 1 value, in spite of a positive noise added at *Q*. However, in conventional 6T SRAM cell, if a positive noise, ΔV_{noise} is added at $Q = 0 + \Delta V_{noise}$ and reaches the threshold voltage of the opposite NMOS transistor, which turns it ON. This makes a discharging path from QB to *GND*, which would flip the state of QB from logic 1 to logic 0 and degrades the RSNM value.

3.5.4 Analysis of PFC8T SRAM

The 32 × 64-bit SRAM array is simulated using 65nm standard CMOS technology. The post-layout simulations in iso-area conditions are carried out to determine various constraints like leakage current, power, read-write delay and power, PDP, RSNM, WSNM, DNM, half-select issues, and WTP. Furthermore, all the constraints are observed at different temperature values ranging from 0°C to 100°C and at different process corners, namely, FF, SS, TT, SF, and FS using MC simulations at standard 6σ variations.

3.5.4.1 Leakage Power Estimation

Leakage current in SRAM is measured as the current drawn from V_{DD} to GND when the SRAM cell is at static or hold condition. The mean (μ) values of leakage power are observed at different temperatures, as shown in Figure 3.33. It is observed that the SRAM provides low leakage power variations at different temperature values. It shows a notable improvement and consumes negligible leakage power as compared with 6T and RD8T SRAM at various PVT values.

3.5.4.2 Read Delay and Power Analysis

Read delay is measured when *RWL* is HIGH and *BLB* discharges and reaches the least sensing offset voltage required of the SA [31]. It is obtained from a 32-bit SRAM cell column architecture having a bitline capacitance of 320fF at the worst-case (SS) PC. The power measured for a successful read is defined as the read power of SRAM. Figure 3.34 shows the plot of the read delay and power in the SS and FF process corners. It is seen that PFC8T SRAM provides similar read access times as related to 6T SRAM with less read power.

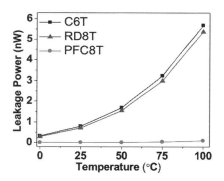

FIGURE 3.33 Leakage power of PFC8T SRAM compared with 6T and RD8T SRAM at various temperature values.

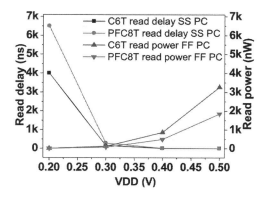

FIGURE 3.34 Read delay and power consumption at different supply voltages.

SRAM Bitcells over Conventional Memories 111

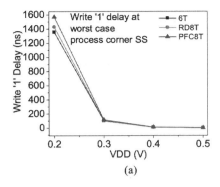
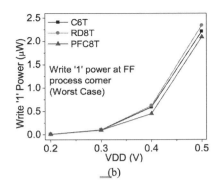

FIGURE 3.35 (a) Write '1' delay, power, and the power delay product in the worst-case PC at different V_{DD}, and (b) the distribution curve of write '1' delay at the SS PC.

3.5.4.3 Write Delay and Power Analysis

The write delay and power are analyzed in a similar way as the previous bitcells. The write '1' access time is measured until storage node Q reaches 0.9 V_{DD}. Similarly, write '0' access time measure until Q discharges and reaches 0.1 V_{DD}. The write dynamic power is measured as the product of the average current flow and the supply voltage at the write access time. Figure 3.35a shows the mean (μ) and sigma (σ) values of write '1' delay and power at various supply voltages. The delay time and power are compared with 6T and RD8T SRAM, which shows the PFC8T SRAM, provides a similar write access time. The write power of PFC8T SRAM shows a reduction of 2.4% and 7% as compared to 6T and RD8T, respectively at 0.3V V_{DD} as shown in Figure 3.35b. Due to a profound reduction in leakage power with a slight improvement in read-write power, the featured bitcell shows a better alternative for power-hungry wireless devices such as mobiles, laptops, and medical portable devices.

3.5.4.4 RSNM and WSNM

The RSNM is measured by applying a DC noise voltage source at one of the storage nodes, Q or QB, and examining its effect on the opposite node. The RSNM is examined in the read operation when *RWL* is HIGH and *WWL* is LOW. The *CS* is kept LOW, which disconnects the path between node Q and *GND* and makes a decoupled logic from *BLB* to *GND*. The RSNM of 8T is shown in Figure 3.36a, which shows an improvement by 2.86× compared to 6T as shown in Figure 3.36b at 0.3V V_{DD}. Simultaneously, the RSNM values for 6T, RD8T, and PFC8T SRAM are observed at different temperature values as shown in Figures 3.37a, 3.37b, and 3.37c, respectively. The comparison of RSNM values is also shown in Figure 3.37d. It shows that the presented bitcell has a better read stability than 6T and RD8T SRAMs. Similarly, the WSNM is considered at the time of the write operation by initiating linear DC noise at one of the storage nodes and observing the effect of the noise at other. To achieve this, the *WWL* is kept HIGH, *RWL* kept LOW, and the *CS* is kept HIGH. Figure 3.36a shows the WSNM of PFC8T, which shows an improvement of 7.1% (128mV) compared to 6T SRAM (119.5mV) as shown in Figure 3.36b.

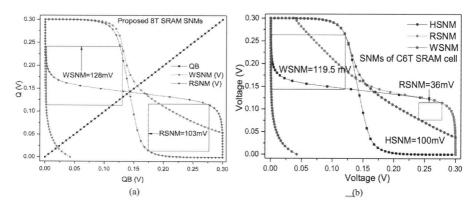

FIGURE 3.36 (a) RSNM and WSNM of 8T SRAM are observed at 0.3V V_{DD} and (b) RSNM, HSNM, and WSNM of 6T SRAM 0.3V V_{DD}.

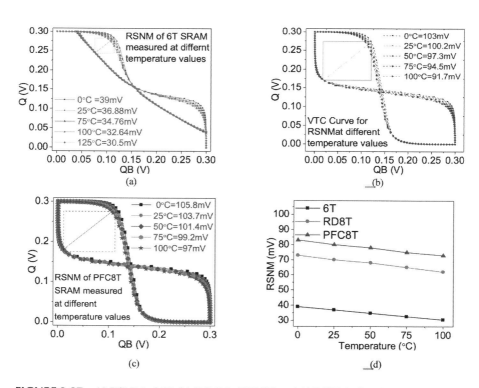

FIGURE 3.37 (a) RSNM of 6T, (b) RSNM of RD8T, and (c) RSNM of PFC8T are observed at 0.3V V_{DD} at different temperature values. (d) Comparison of RSNM values of 6T, RD8T, and PFC8T at different temperatures.

SRAM Bitcells over Conventional Memories   113

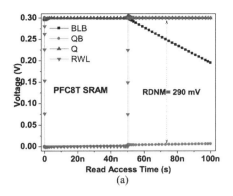 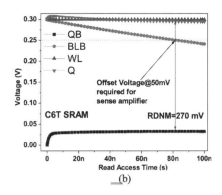

FIGURE 3.38 (a) RDNM of PFC8T SRAM and (b) RDNM of 6T SRAM at 0.3V V_{DD}.

3.5.4.5 Dynamic Read Margin

To measure dynamic read margin, *RWL* is activated HIGH and *WWL* is kept at LOW. Furthermore, when the *BLB* reaches the minimum offset voltage required for sensing the stored info, the difference between the values of storage nodes Q and QB is defined as the read dynamic noise margin (RDNM). Figures 3.38a and 3.38b show the observation of RDNM of 8T SRAM. The RDNM of PFC10T SRAM is 7.4% better than 6T SRAM at the worst-case process corner.

3.5.4.6 WTP

The WTP is measured at the time of write operation while *WWL* asserted HIGH and *RWL* is LOW. It can be observed by two methods: one by adding a linearly varying DC voltage source at *BL* and observing its effect on the *BLB* and the other by varying *WWL* and writing through *BL* and *BLB*. The WTP, when *WWL* varies comes out to be 30mV which is 43% better than 6T SRAM (21mV) at 0.3V V_{DD} in the worst-case (SF) PC. Figure 3.39 shows the comparison of simulation results of WTP observed for PFC8T, 6T, and RD8T at different supply voltages. The WTP is an important factor for improving the writing ability of an SRAM.

3.5.5 Results Summary of 2-kb Array

A read decoupled PFC 8T SRAM cell for fast speed and low power object tracking is presented. Furthermore, the bitcell is used to implement a 2-kb macro. The bitcell is designed in 65nm standard CMOS technology. It is observed from the simulation results that the bitcell shows superior performance in terms of cell stability, WTP and leakage power as compared to existing SRAM cells. The bitcell also shows a better ON-to-OFF current ratio (I_{on}/I_{off}) and energy for read and write operations. Furthermore, the leakage power and energy are observed for the 2-kb macroblock, which shows improved outcomes at different supply voltages in the subthreshold regime. The power consumption of 8T SRAM is compared with the total power consumed by 6T and RD8T SRAM cells. It is observed that the total power is reduced by

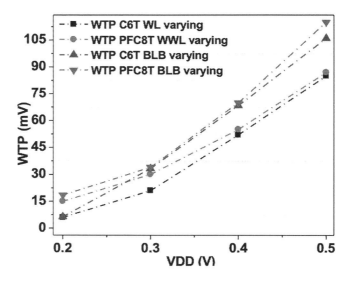

FIGURE 3.39 WTP at the worst-case (SF) PC.

TABLE 3.11
Summary of Mean (μ) Values of Simulation Results of PFC8T SRAM–Based 2-kb Macro in 65nm Technology

SRAM	Leakage power (pW)	Write '1' delay (ns)	Write '1' power (nW)	Read delay (ns)	Read power (nW)	RSNM (mV)	WSNM (mV)	WTP (mV)
6T	850	99.43	98.4	9.17	860.4	36	119.5	21
RD8T	786	104	103.1	10.8	720	100	120	30
PFC8T	10.5	115.8	96	17	492	103.7	128	28

1/3× compared to 6T and RD8T SRAM cells. Table 3.11 shows the summary of post-layout simulation results of PFC8T SRAM–cell-based 2-kb SRAM at 0.3V power supply. Furthermore, it is observed that the WSNM is improved by 7.1%, RSNM by 2.86×; WTP by 43% compared to 6T at the worst-case PCs. The PDP for write '1' operation is reduced by 32.85% compared to a 6T SRAM cell. The leakage power is also reduced by 82× as compared to 6T SRAM–based 2-kb array at 0.3V V_{DD}.

3.5.6 Application of PFC8T SRAM on an Object Tracking System

The working principle of subthreshold 8T SRAM and simulation results of 2-kb SRAM at different PVT values is stated in the previous section. The objective of the 8T SRAM is to construct a cache memory for a high-stability (high-accuracy)

SRAM Bitcells over Conventional Memories

system that operates on battery power. The high read stability and ULP operation in the subthreshold region arise as a convenient choice for designing SRAM array architecture for image/video processing systems like object detection and tracking. The image/video processing for object detection and tracking requires a high-speed, low leakage power, and high-accuracy digital system for constructing an effective yield. Moreover, the memory required for storing and accessing bits needs to be very fast. Therefore, an SRAM is chosen as a preference to develop a very high-speed object tracking memory system. The block diagram of a modified approach of memory reduction in object detection and tracking is explained in Figure 3.40. Also, minimizing the memory size for object detection and tracking an algorithm is presented. The algorithm uses a predefined rectangular box called a macroblock, selected at the probable entries of the object where the probability of object sighting is high. This helps reduce the memory required to detect and track the object by differencing the present macroblock with the previous one instead of differentiating the whole frame. The modified algorithm cuts the memory size by a huge factor as explained in the succeeding section. Later, the PFC8T SRAM is used to reduce the overall system and leakage power with better accuracy in subsequent sections.

3.5.6.1 Object Detection and Tracking

An image is an array represented by a number of bits. An image is defined as a two-dimensional function $f(X, Y)$, where X and Y are the coordinates of the image, and the amplitude of $f(X, Y)$ defines the intensity of the image at that point. The object detection and tracking are achieved using various steps, which are mentioned in the block diagram shown in Figure 3.40.

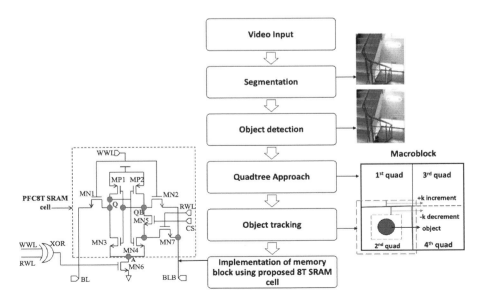

FIGURE 3.40 Block diagram of a modified object tracking algorithm.

In the object detection and tracking algorithm, rectangular shape macroblocks are placed at every entrance point in the field of view (FOV). Furthermore, the object is detected by considering the difference of the root-mean-square (RMS) values of reference and the current frame. After detecting an object, the quadtree-based approach is employed to minimize the bounding box. This is one of the major contributions of the research to reduce the processing time, logical comparators, memory utilization, and henceforth power consumption. Thereafter, object tracking is achieved by using macroblock resizing. Furthermore, the implementation of the memory required for the approach is performed using a ULP PFC8T SRAM cell.

3.5.6.2 Segmentation

Image segmentation is a basic step in image processing and is also a significant part of image analysis. The macroblocks are placed at various locations of the FOV of the camera, where the probability of an object entering is high. These entry blocks, which are rectangular in shape, are denoted as initial macroblocks. The captured red green blue (RGB) frame is converted into a gray image and the initial macroblocks are considered as reference macroblocks. Figure 3.41a and Figure 3.41b shows the selected macroblocks at the entry sites and the object detection at the initial macroblock, respectively.

3.5.6.3 Object Detection

To detect the object, RMS values of macroblocks are considered. The RMS value of the macroblock is calculated using Equation 3.1.

$$rms = \sqrt{\frac{\sum_{k=1}^{N} p_k^2}{N}}, \qquad (3.1)$$

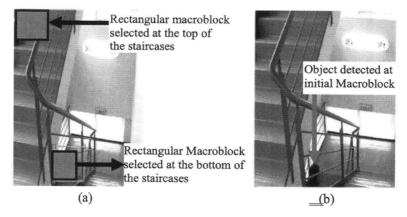

FIGURE 3.41 (a) Rectangular macroblocks selected at the entry points in the FOV, and (b) object detected at the initial macroblock.

SRAM Bitcells over Conventional Memories

where *rms* denotes the RMS value of the macroblock, N is the number of pixels in the macroblock, and p_k is gray intensity value of the kth pixel. The difference between RMS values of the reference macroblock and the current macroblock is determined and compared with an adaptive threshold value for object detection. When the difference in RMS values becomes greater than or equal to the threshold, the object is detected in the FOV.

3.5.6.4 Quadtree Method

To determine the exact location of an object in the macroblock and reduce the size of the bounding box, a quadtree decomposition approach is used [32]. The macroblocks of reference and current frames are divided into four quadrants. The RMS values of corresponding quadrants of reference and current frames are compared. If the difference of RMS values of at least three quadrants is greater than the threshold, then the minimum bounding box is achieved. Otherwise, the quadrants with a difference in RMS values greater than the threshold value will be further divided into four parts. The procedure is repeated until the minimum bounding box is achieved. It should be noted that if we divide the macroblock into four parts, then the new threshold of each quarter is changed to 2× of the threshold value of initial macroblock, according to Equation 3.1. The number of pixels after dividing the frame into four parts will be N/4; thus, the RMS becomes twice the original value. In Figure 3.42a, the quadtree approach is defined by using the flow diagram.

Figure 3.42b shows the initial macroblock, which is further divided into four equal quarters. As shown in Figure 3.42b, if the difference of RMS values of the

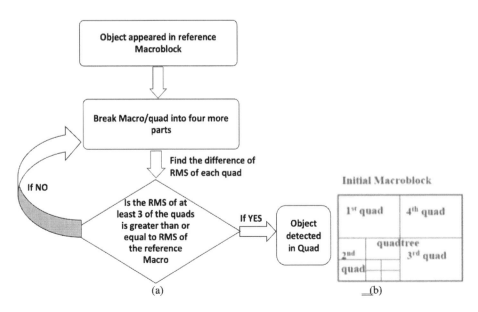

FIGURE 3.42 (a) Quadtree method for reducing the macroblock size and (b) macroblock divided into four quarters to reduce the bounding box.

second quad become greater than or equal to the 2× the threshold value of the initial macroblock, then it would be further divided into four parts. It will go on until at least three quarters reach $\sqrt{2^{(i+1)}}$ × the threshold of the initial macroblock, where 'i' represents the number of divisions performed for each quadrant.

3.5.6.5 Tracking Using Macroblock Resizing

After minimizing the bounding box around the object, our next objective is to improve the tracking speed and area utilization in terms of memory. The lower memory usage consequently results in minimizing the power consumption of the memory blocks. Figure 3.43 shows the macroblock where object is detected in the second quad. After applying the quadtree method, the size of the quad is reduced to 1/4× the size of the original macroblock. Therefore, to improve the tracking speed with a high accuracy rate and less memory utilization, it is intended to have motion vectors in all four directions as shown in Figure 3.43a and Figure 3.43b. In Figure 3.43, k is the directional vector pixels (DVPs) taken outside and inside of the rectangular macroblock or the sub-blocks reckoned from the quadtree approach.

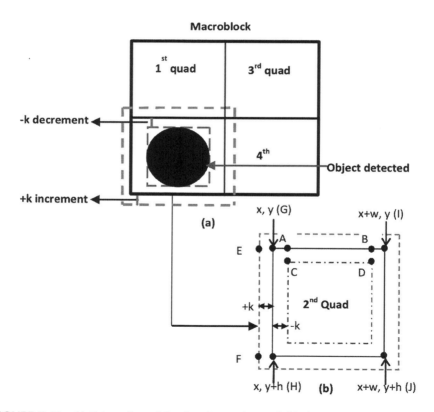

FIGURE 3.43 (a) Selected quad for the observation and (b) the selected quad where an object is detected.

Further, for tracking an object or group of objects in a scene, an object detection/tracking system is required. The processing unit comprises a memory block, comparators, counters, decoders, and a clocking unit, as shown in Figure 3.44. The tracking system shown in the figure detects the motion vectors of consecutive frames. The memory block is required to store current and reference macroblocks. Furthermore, an 8-bit comparator is used to compare the change in information of the macroblocks. Since an SRAM is fast and consumes less power, it is used as a memory device to achieve a fast and low-power tracking system. Evidently, the cache memory block for object detection/tracking requires arrays of the SRAM cell to write, store, compare, and read binary information [33–35]. According to Hori and Kuroda [32], a micro-programmable real-time video signal processor (VSP)–large-scale integration (LSI) has been developed for constructing a parallel video signal processing system. The VSP-LSI utilizes a multistage pipelined architecture and can handle complex image and signal processing applications such as high-speed edge detection, face detection, and motion compensation. The LSI contains an SRAM-based cache for the realization.

Hori and Kuroda [34] fabricated a real-time face detection system using 0.13-μm CMOS technology. It consists of 75,000 gate logic, 58,000-bit SRAM, and Advanced RISC Machine (ARM)-advanced microcontroller bus architecture (AMBA) interface. In [35], a dynamically reconfigurable SRAM array for low-power mobile multimedia applications is displayed, which shows how an SRAM array can significantly change the overall multimedia power. However, a large number of memory arrays in portable tracking devices consume a huge amount of processing and leakage power.

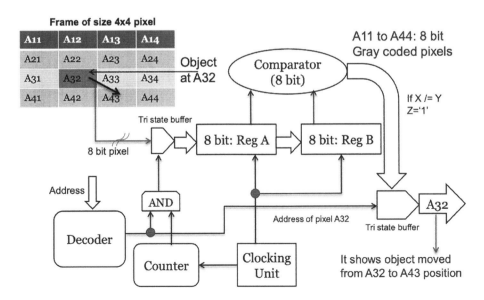

FIGURE 3.44 The object tracking memory block for storing and comparing the gray-coded frames.

Therefore, a subthreshold 8T SRAM cell is presented here to reduce leakage power consumption, which is further utilized to implement SRAM arrays for a tracking device.

3.5.7 Object Tracking Analysis and SRAM Utilization: Results and Comparisons

The object tracking algorithm uses a smaller number of memory blocks to compare two different frames as shown in Table 3.12. Instead of comparing the whole frame of size 704 × 576 in the given example, the algorithm uses a macroblock-based approach. The average memory utilized in this case can be determined from the mean of the total number of pixels used to track the object. To measure the average power consumption of the memory block, the amount of power consumption by a memory bit cell is estimated. To authenticate the algorithm in MATLAB tool, a monocular camera–based surveillance video is considered. To support high-level video surveillance tasks and other processing algorithms, the server is equipped with an INTEL Core i7 4710HQ 2.5GHz CPU and 8GB RAM. The average leakage power/frame for 6T, RD8T, and PFC8T SRAM is equivalent to 272nW, 251.5nW, and 3.2nW, respectively, as shown in Table 3.13. The average total power consumption is the sum of power consumed while writing, reading, and holding (leakage) data into the SRAM for tracking an object. The total power consumed is equivalent to 38.33mW, 33mW, and 23.5mW for 6T SRAM, RD8T, and PFC8T SRAM–based memory array required for tracking an object. Figure 3.45 shows the set of frames extracted from a video. In the first frame of the video, the macroblock of size 64 × 79

TABLE 3.12
Average Memory Required/Frame for Object Detection and Tracking along with the Object Detection and Tracking Delay

Macroblock size (pixels)	Memory required for object detection/frame (kb) without quadtree pixelation	Memory size required after quadtree pixelation (kb)	Average memory required for tracking/frame (kb)	Delay in detection (s)	Average delay in tracking/frame (s)
13 × 16	0.20	0.06	37.15	2.08	0.15
23 × 18	0.41	0.12	39.85	1.73	0.15
73 × 56	4.08	1.00	39.65	1.71	0.15
64 × 79	5.00	1.30	36.84	2.12	0.18
75 × 74	5.55	5.55	39.00	1.84	0.16
100 × 112	11.2	2.85	38.20	1.80	0.17
115 × 119	13.6	3.40	42.37	1.70	0.15
149 × 148	22.0	10.0	41.40	2.00	0.16
215 × 137	29.4	7.45	37.00	1.74	0.16

TABLE 3.13
Average Leakage Power Observed in Tracking an Object at 0.3V V_{DD}

SRAM	Average leakage power (nW)	Average total power (mW) at worst-case (FF) PC
6T	272nW	38.33mW
RD8T	251.5nW	33mW
PFC8T	3.2nW	23.5mW

FIGURE 3.45 Object tracking using quadtree pixelation where the bounding box is selected at the entry points of the FOV.

is selected. To detect the object in the scene, a threshold value is determined, which is equal to the peak of the difference of RMS values of the selected macroblocks. In Figure 3.45, it is shown that the object enters the scene at frame 129, and after that, a quadtree approach is used to reduce the bounding box size. Consequently, by using the quadtree approach, the object can be tracked by using one fourth or one eighth the number of pixels as compared to the original macroblock. The object tracking output using macroblock resizing is shown in Figure 3.45. Table 3.14 shows the accuracy rate of the object tracking algorithm and other existing works. The accuracy rate is calculated by using Equation 3.2, where the percentage of difference in the area of the actual bounding box to the estimated bounding box is measured. In the

TABLE 3.14
Comparison of Tracking Accuracy Rate with Respect to the Actual Bounding Box

Tracking accuracy rate (%)	Classical background difference [27]	Background difference [28]	Fusion algorithm [28]	PETS 2009 dataset [29]	iLIDS medium sequence [29]	Tracking using macroblock resizing
	88.5%	87.4%	91.0%	96.2%	95.2%	96.5%

present case, the accuracy rate comes out to be 96.5%. The accuracy rate could be different for different macroblock sizes.

Tracking Accuracy Rate (TAR in %) =

$$\sum_{i=k}^{N} \frac{Area\ of\ estimated\ bounding\ box - Area\ of\ actual\ bounding\ box}{Area\ of\ actual\ bounding\ box}, \quad (3.2)$$

where k is the frame where the object is detected and N is the total number of frames present in the video.

REFERENCES

1. Do Anh-Tuan et al. (2008), Hybrid-mode SRAM sense amplifiers: New approach on transistor sizing, *Circuits and Systems II*, vol. 55, no. 10, pp. 986–990.
2. Z. Bo et al. (2008), A variation-tolerant sub-0.2V 6T the subthreshold SRAM, *IEEE Journal of Solid-State Circuits*, vol. 43, no. 10, pp. 2338–2348.
3. B. H. Calhoun and A. P. Chandrakasan (2007), A 256-kb 65-nm sub-threshold SRAM design for ultra-low-voltage operation, *IEEE Journal of Solid-State Circuits*, vol. 42, no. 3, pp. 680–688.
4. S. W. Sun and P. G. Y. Tsui (1995), Limitations of CMOS supply-voltage scaling by MOSFET threshold-voltage variation, *IEEE Journal of Solid-State Circuits*, vol. 30, no. 8, pp. 947–949.
5. T. H. Kim et al. (2007), An 8T the subthreshold SRAM cell utilizing reverse short channel effect for write margin and read performance improvement, *IEEE Custom Integrated Circuits Conference*, San Jose, CA, USA, pp. 241–244.
6. K. Zhang et al. (2004), SRAM design on 65 nm CMOS technology with integrated leakage scheme, Symp. *On Digest of Technical Papers on VLSI Circuits*, Honolulu, HI, USA, pp. 294–295.
7. J. Chen et al. (2006), An ultra-low-power memory with a subthreshold power supply voltage, *IEEE Journal of Solid-State Circuits*, vol. 41, no. 10, pp. 2344–2353.
8. L. Chang at al. (2008), An 8T-SRAM for variability tolerance and low-voltage operation in high-performance caches, *IEEE Journal of Solid-State Circuits*, vol. 43, no. 4, pp. 956–963.
9. J. P. Kulkarni and K. Roy (2012), Ultra low-voltage process-variation-tolerant Schmitt-trigger-based SRAM design, *IEEE Transactions on Very Large-Scale Integration System*, vol. 20, no. 2, pp. 319–332.

10. S. Tawfik and V. Kursun (2008), Low power and robust 7T dual-VT SRAM circuit, *IEEE International Symposium on Circuits and Systems*, Seattle, USA, pp. 1452–1455.
11. M. F. Chang et al. (2010), A differential data-aware power-supplied (D2AP) 8T SRAM cell with expanded write/read stabilities for lower VDDmin applications, *IEEE Journal of Solid-State Circuits*, vol. 45, no. 6, pp. 1234–1244.
12. S. Pal and A. Islam (2016), Variation tolerant differential 8T SRAM cell for ultralow power applications, *IEEE Transactions on Computer-Aided Design of Integrated Circuits and Systems*, vol. 35, no. 4, pp. 549–558.
13. C. B. Kushwah and S. K. Vishvakarma (2016), A single-ended with dynamic feedback control 8T the subthreshold SRAM cell, *IEEE Transactions on Very Large-Scale Integration Systems*, vol. 24, no. 1, pp. 373–377.
14. M. F. Chang et al. (2011), A 130mV SRAM with expanded write and read margins for the subthreshold applications, *IEEE Journal of Solid-State Circuits*, vol. 46, no. 2, pp. 520–529.
15. A. A. Mazreah and M. T. M. Shalmani (2011), New configuration memory cells for FPGA in nano-scaled CMOS technology, *Elsevier Microelectronics Journal*, vol. 42, no. 11, pp. 1187–1207.
16. A. A. Mazreah and M. T. M. Shalmani (2012), Low-leakage soft error tolerant port-less configuration memory cells for FPGAs, *Integration the VLSI Journal*, vol. 46, no. 4, pp. 413–426.
17. N. Verma and Anantha P. Chandrakasan (2008), A 256-kb 65 nm 8T the subthreshold SRAM employing sense amplifier redundancy, *IEEE Journal of Solid-State Circuits*, vol. 43, no. 1, pp. 141–149.
18. Ya-Chun Lai and Shi-Yu Huang (2008), A resilient and power-efficient automatic-power-down sense amplifier for SRAM design, *Circuits and Systems II: Express Briefs*, vol. 55, no. 10, pp. 1031–1035.
19. B. S. Reniwal et al. (2015), Ultra-fast current mode sense amplifier for small ICELL SRAM in FinFET with improved offset tolerance, *Springer Circuits Systems and Signal Processing*, pp. 1–20.
20. M. H. Chang et al. (2011), A 1-kb 9T the subthreshold SRAM with bit-interleaving scheme in 65 nm CMOS, *IEEE/ACM International Symposium on Low Power Electronics and Design*, Fukuoka, pp. 291–296.
21. S. Pal and A. Islam (2016), 9T SRAM cell for reliable ultralow-power applications and solving Multibit soft-error issue, *IEEE Transactions on Device and Materials Reliability*, vol. 16, no. 2, pp. 172–182.
22. A. Islam and M. Hasan (2012), Leakage characterization of 10T SRAM cell, *IEEE Transactions on Electron Devices*, vol. 59, no. 3, pp. 631–638.
23. B. S. Reniwal et al. (2015), Dataline isolated differential current feed/mode sense amplifier for small icell SRAM using FinFET, *25th Great Lakes Symposium on VLSI (GLSVLSI)*, Pittsburgh, PA, USA, pp. 95–99.
24. S. Mukhopadhyay, A. Raychowdhury and K. Roy (2003), Accurate estimation of total leakage current in scaled CMOS logic circuits based on compact current modeling, *Proceedings of the 40th Annual Design Automation Conference*, Anaheim, CA, USA, pp. 169–174.
25. H. Qin et al. (2004), SRAM leakage suppression by minimizing standby supply voltage, *5th International Symposium on Quality Electronic Design*, San Jose, CA, USA, pp. 55–60.
26. R. K. Behera et al. (2012), Multi-camera-based surveillance system, *World Congress on Information and Communication Technologies (WICT)*, Trivandrum, pp. 102–108.
27. C. Stauffer and W. E. L. Grimson (2000), Learning patterns of activity using real-time tracking, *IEEE Transactions on Pattern Analysis and Machine Intelligence*, vol. 22, no. 8, pp. 747–757.

28. G. Han, X. Li, N. Sun and J. Liu (2014), A robust object detection algorithm based on background difference and LK optical flow, *International Conference on Fuzzy Systems and Knowledge Discovery (FSKD)*, Xiamen, pp. 554–559.
29. W. Wang, R. Nevatia and B. Yang (2015), Beyond pedestrians: A hybrid approach of tracking multiple articulating humans, *IEEE Winter Conference on Applications of Computer Vision*, Waikoloa, HI, pp. 132–139.
30. E. Grossar et al. (2006), Read stability and write-ability analysis of SRAM cells for nanometer technologies, *IEEE Journal of Solid-State Circuits*, vol. 41, no. 11, pp. 2577–2588.
31. B. S. Reniwal, P. Bhatia and S. K. Vishvakarma (2017), Design and investigation of variability aware sense amplifier for low power, high-speed SRAM, *Microelectronics Journal*, vol. 59, pp. 22–32.
32. E. Shusterman and M. Feder (1994), Image compression via improved quadtree decomposition algorithms, *IEEE Transactions on Image Processing*, vol. 3, no. 2, pp. 207–215.
33. M. Yamashina et al. (1987), A micro-programmable real-time video signal processor (VSP) LSI, *IEEE Journal of Solid-State Circuits*, vol. 22. no. 6, pp. 1117–1123.
34. Y. Hori and T. Kuroda (2007), A 0.79mm2 29-mW real-time face detection core, *IEEE Journal of Solid-State Circuits*, vol. 42, no. 4, pp. 790–797.
35. M. Cho, J. Schlessman, W. Wolf and S. Mukhopadhyay (2011), Reconfigurable SRAM architecture with spatial voltage scaling for low power mobile multimedia applications, *IEEE Transactions on Very Large-Scale Integration (VLSI) Systems*, vol. 19, no. 1, pp. 161–165.

4 Offset Correction in the Sense Amplifier

4.1 INTRODUCTION

A typical simplified static random-access memory (SRAM) architecture, along with a simple schematic illustration of column circuitry, is shown in Figure 4.1. The data storage element, or core, consists of individual 6-transistor (6T) memory cells arranged in an array of horizontal rows and vertical columns. Here each cell is capable of storing one bit of binary information. Also, each 6T cell shares a common connection with the other cells in the same row called wordline (WL) and another common signal with the other cells in the same column called column select (CS). In this structure, there are 2^n rows and 2^m columns. Thus, the total number of memory cells in this array is $2^m \times 2^n$. The denser memory integrated circuits (ICs) have many such blocks depending on the memory size and preferred architecture. The 6T SRAM cell contains a pair of weak cross-coupled inverters (M0–M3) holding the state and a pair of access transistors (M4–M5) to read or write the state. The positive feedback corrects the disturbances caused by leakage or noise.

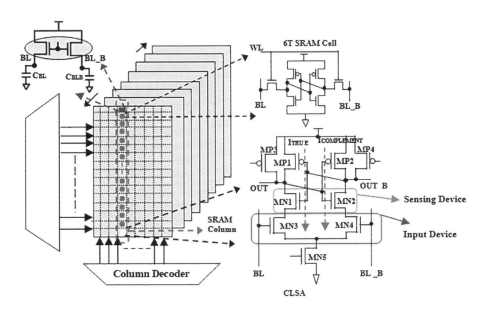

FIGURE 4.1 Illustration of implemented SRAM architecture with current latch sense amplifier.

DOI: 10.1201/9781003213451-4

To access a particular memory cell, that is, a data bit in this array, the corresponding CS and the corresponding WL must be activated. The row and column selection operations are accomplished by row and column decoders, respectively.

4.2 DATA SENSING METHODS IN EMBEDDED MEMORIES

Bitline sensing can be classified as large-signal sensing or small-signal sensing. Figure 4.2 depicts the schematic of large and small signal sensing in SRAM. In large-signal, also known as single-ended, sensing, one of the bitlines directly acts as an input to the sense amplifier and bitline swing between V_{DD} and GND just like an ordinary digital signal [1]. The high-skew inverter is an example of large-signal sensing. To reduce the parasitic delay, the bitline can be hierarchically divided into multiple local bitlines and then combined to drive a global WL. Large-signal sensing consumes much power due to large voltage swings. In small-signal sensing, also called differential sensing, both bitlines act as an input to the sense amplifier [2]. Now, once WL is asserted, one of the two bitlines changes by a small amount. A sense amplifier detects the small differential voltage and produces a full swing output. This saves the delay of waiting for a full bitline swing and reduces energy consumption if the bitline swing is terminated after sensing. However, the array requires a timing circuit to indicate when the sense amplifier should fire, and if the time is too short, the wrong answer may be sensed.

It is much clearer that sense amplifier plays an important role in SRAM designs and affects the speed, overall power, and read failure rate for SRAM macros. The purpose of the sense amplifier is to sense and amplify a small voltage difference between the two input nodes, *BL* and *BLB*, which prevents a full-swing discharge on

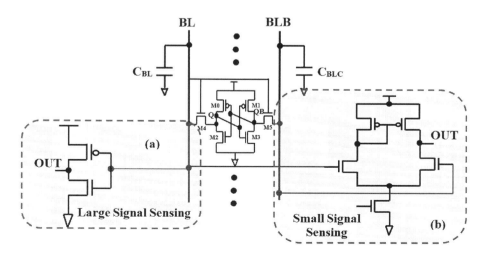

FIGURE 4.2 Sensing techniques: (a) Large-signal sensing. (b) Small-signal sensing.

the aforesaid bitlines and hence improves the cache access latency and reduces the dynamic power consumption. Shekhar Borkar, Intel fellow, suggests that

> [s]ense amplifiers are not tractable in future and will go away but not in memories https://scholarworks.umass.edu/cgi/viewcontent.cgi?article=1253&context=theses.

The sense amplifiers that are used in SRAM are mainly differential in nature. The differential sense amplifiers can distinguish smaller signals from noise due to their high common mode rejection ratio providing good reliability. Different types of sense amplifiers have been proposed to improve the read latency, sensing delay, and memory failure probability at a given supply voltage. The typical classification approaches are discussed in the following.

4.2.1 Voltage Mode Approach

In conventional memories, voltage-mode sense amplifiers (VMSAs) are used that offer high input impedance to the bitlines. This allows the sense amplifier to provide a high voltage gain with the use of simple circuits. The voltage mode sense amplifier requires differential discharging of the bitline capacitance for sensing the voltage difference. It activates after a certain amount of differential voltage is developed on the bitlines. Here, the time to develop a certain differential voltage to appear depends on the bitline capacitance, which will increase with the increase in the bitline capacitance (i.e., number of cells in the column) and increased leakage current. This problem will worsen with the coming technology generation. Figure 4.3a shows the conventional latch-type VMSA that is used to read the data from several types of memories since these types of sense amplifier achieve fast decisions due to strong positive feedback. In a VMSA, two cross-coupled inverters provide positive feedback. The sense enable signal (SEN) turns on the amplifier and starts the sensing operation. Depending on the polarity of the voltage difference between the nodes SO and SOB, the sense amplifier will flip in one or the other direction. Switching SEN to logic low resets the latch before the next read can start. Figure 4.3b portrays the post-layout circuit characteristics with inter-die and intra-die variation obtained by Monte Carlo simulation of 10,000 iterations. It can be observed that out of the 10,000 samples, latch type SA suffers from lower functional yield and higher failure probability. This shows the sense amplifier performance degrades under process variation. To enable the sense amplifier design less vulnerable to variations, bitline differential should be greater than the sense amplifier offset voltage $(\Delta V_{BL} > V_{OS})$. Therefore, aerospace, medical, and other high-reliability applications call for a design that can offer a high-speed solution to the offset problem.

Furthermore, nodes SO and SOB are input and output terminals at the same time. Therefore, the circuit cannot be connected directly to the bitline; otherwise, the circuit would attempt to discharge the bitline capacitance during the decision phase and would increase delay and power. A solution is either to decouple the bitline by a multiplexer or to use pass gates, forming a decoupling resistor. Both devices cause a voltage drop that deteriorates the available input voltage difference and affects the speed and reliability of the sense amplifier. This issue was solved by Kobayashi et al. [3] by adding two additional transistors (M6 and M7; Figure 4.4), which offer high

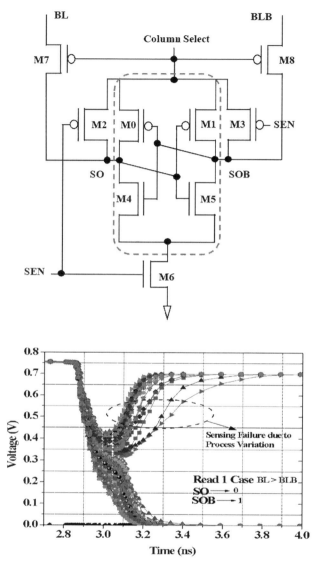

FIGURE 4.3 Schematic illustration of (a) VMSA and (b) output characteristic obtained by Monte Carlo simulation for 10,000 iterations for VMSA.

input resistance. The current flow through the differential input transistors M6 and M7 controls the serially connected latch; therefore, it is called the current latch sense amplifier (CLSA) as depicted in Figure 4.4.

Here, before sensing, output terminals SO and SOB are precharged to V_{DD} by the transistors M2 and M3. The SEN signal starts the sensing operation by turning on

Offset Correction in the Sense Amplifier

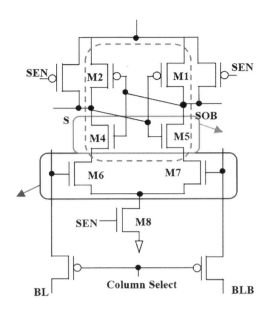

FIGURE 4.4 Schematic illustration of current latch sense amplifier.

M8. Since M6 and M7 are connected to the bitlines, the drain currents of M6 and M7 start to discharge to output nodes SO and SOB. Depending on the bitlines voltage, the drain current of M6 and M7 will be different and cause different discharging speeds at SO and SOB. Since the output nodes are precharged to V_{DD} before reading, both p-channel transistors M0 and M1 remain turned off until the output voltage reaches $V_{DD} - V_{THP}$. Once the node either SO/SOB (depending on memory data) reaches $V_{DD} - V_{THP}$, it turns on the p-type metal-oxide semiconductor (PMOS) device (M0 or M1), and strong positive feedback or latching action starts to generate the output. Voltage mode implementations are popular among researchers, which also do have some limitations. The input offset of the sense amplifier in voltage mode sets the higher level for the required bitline discharge, thereby increasing the energy consumption. This offset also affects the sensing delay or even the functionality of the circuit, depending on the extent of the process variation. Therefore, determining the worst-case possibility of process variation is highly significant in sense amplifier design. The minimal target value of the required bitline discharge depends on the technology, sense amplifier design, sizing, and target yield level.

4.2.2 Current Mode Approach

Like other integrated circuits today, complementary metal-oxide semiconductor (CMOS) memories are required to increase speed, improve capacity, and maintain low power dissipation. These objectives are somewhat conflicting when it comes to sense amplifiers in memories. Although the components become faster when evolving to

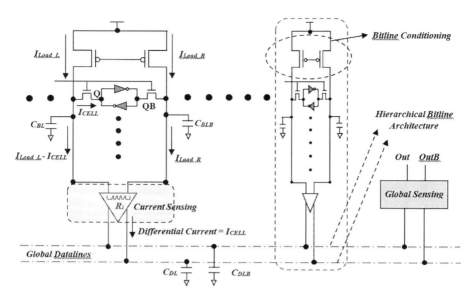

FIGURE 4.5 Schematic illustrating the current mode sensing with SRAM memory column.

deep submicron technologies, signal delay over long interconnects constitutes a major bottleneck in achieving high circuit speed. The increased memory capacity usually results in enhanced bitline parasitic capacitance. This increased bitline capacitance, in turn, slows down voltage sensing and makes bitline capacitance swing and energy-expensive, resulting in slower and more energy-hungry memories. Therefore, in denser memories, bitline capacitance is a severe design bottleneck. It shows that a solution to this problem will have to involve drastically reduced on-chip voltage signal swings.

The current mode sensing employs the use of current signals with hardly any voltage swing to diminish the speed bottlenecks caused by interconnect delay in CMOS very large scale integration (VLSI) circuits. Figure 4.5 illustrates the typical current mode sensing architecture in static-random access memory (SRAM). The gist of current mode sensing is that it offers very low input resistance and instantaneously transfers the differential current to global bitlines without waiting for large bitline swing at highly capacitive bitlines. This results in reduced sensing delay and power consumption in data sensing.

4.3 DESIGN METRICS FOR EFFICIENT SENSING AMPLIFIER DESIGN

4.3.1 SENSING DELAY AND READ LATENCY

In sense amplifiers, time to sense (T_{sen}) of the sense amplifier is defined as the time required to develop the viable ΔV_{BL} before SEN is enabled. Thus, time to sense dominates the total delay due to the slow discharging process of heavily loaded bit-lines. Once the sense amplifier amplifies the bitline differential and produces a correct output, any further development of the bitline differential signal is redundant and consumes power. Therefore, the higher the offset, the higher the required bitline differential, which results in slower memories. Furthermore, sensing delay is defined as the time between the 50% rise of the SEN signal to the time when the node, which is fully discharging reaches 50% of V_{DD}.

4.3.2 INPUT-REFERRED OFFSET IN SENSING AMPLIFIERS

The minimum voltage difference between bit-lines that the sense amplifier can sense reliably is called its input-referred offset (V_{OS}). Thus, the ability of the sense amplifier to sense small input differences is limited by its input-referred offset. This small voltage difference depends on the WL activation window, which creates the voltage swing at heavily capacitive bitlines. If the SA-SEN signal is asserted early, the sense amplifier cannot amplify the small voltage difference correctly. The overhead of access time and power consumption is increased if the SEN is asserted late. Therefore, the optimum timing for the SEN signal is critical for a high-speed and low-power SRAM device. Thus, the higher the offset of the sense amplifier is, the higher the required differential voltage will be, which results in higher bitline swing (more power consumption) and delayed SEN enabling (slow memory). The sense amplifier offset typically comes due to transistor mismatch in ostensibly identically matched transistor pairs in the sense amplifier, if the mismatch is higher, this results in a higher offset. Figure 4.6 illustrates the important speed, power, and offset correlation for embedded memory design.

High V_{OS} leads to considerable timing and energy overheads as a result of swinging high capacitance bitlines. For high resolution, it is important to minimize the sense amplifier's input-referred offset voltage V_{OS}, which is largely

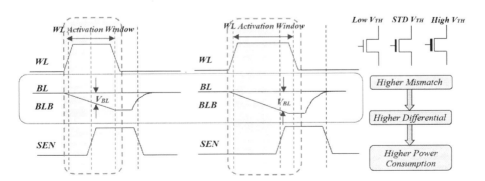

FIGURE 4.6 Schematic illustrating the significance of SEN timing for achieving low power and high speed along with the impact of device mismatch on memory performance and power.

determined by a threshold mismatch of the sensing transistors [4]. The simplest way to reduce the sense amplifier's V_{OS} is to increase the size of the devices [5]. When the required input swing is reduced, a traditional sense amplifier must be made larger and its dynamic energy consumption increases. However, increasing the device size to cope with the sense amplifier offset is also not a preferable choice in lower technology nodes because it resulted in self-loading at the latching nodes (large device capacitance) and makes the sense amplifier slower. However, as the input-referred offset reduces, the required differential voltage on bitlines (*BL*s) to provide 100% yield also reduces. Since the voltage swing is reduced on high capacitance BLs for low offset SAs, the memory consumes lower power than in the case of CLSA. This trade-off between memory yield and power-delay product is thus a major challenge at hand in designing SRAMs. Aerospace, medical, and other high-reliability applications call for a design that can offer a high-speed solution to the offset problem. One approach is to add devices to provide a feedback mechanism to reduce the sense amplifier's sensitivity to V_{th} mismatches.

4.3.3 READ YIELD ISSUE

Aggressive scaling of process technologies driven by Moore's law results in statistical process variations in the transistor parameters such as threshold voltage, channel length, and mobility [6]. Intra-die variations cause differences in transistor characteristics across a single die. In addition, distribution the of device parameters changes from die to die within a wafer and across multiple wafers is called the inter-die variation. Due to intra-die variation, a mismatch occurs. This mismatch in ostensibly identically matched transistor pairs causes the sense amplifier structure to become asymmetrical. The root causes may be many (geometry variations, random dopant fluctuations, oxide thickness variations, edge roughness, etc.), but in any case, this imbalance can be represented by an input offset voltage (V_{OS}). This is the voltage difference that has to be applied to force the cross-coupled inverters to get into metastability. Only for V_{BL} larger than this offset, flips the sense amplifier in the right direction. Otherwise, sensing failure occurs. The sense amplifier's V_{OS} mainly arises from mismatches in the gain factor, the drain current, the threshold voltage, and the layout of the devices used in the sense amplifier [7,8]. Among these contributors, V_{th} mismatch has been identified as the dominant contributing factor to large V_{OS}. In consequence, this larger V_{os} results in higher power consumption and large sensing delays. This brings us to the typical trade-off between memory yield and power-delay product as shown in Figure 4.7. Therefore, in order to adhere to intense scaling trends, sense amplifier design is also highly constrained, especially in the face of emerging limitations ranging from device-level variability to SRAM array-level power consumption. In view of the preceding discussion, developing a sense amplifier with greater offset tolerance is a prerequisite to achieving robust, low power and faster SRAM.

Offset Correction in the Sense Amplifier

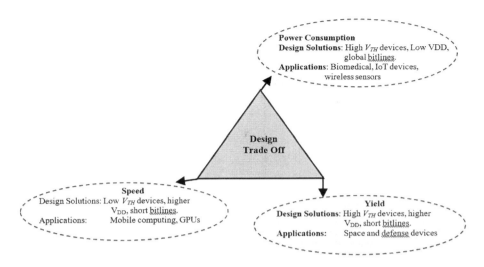

FIGURE 4.7 Design trade-offs in designing SRAM in deep submicron technology nodes.

4.4 VARIABILITY CONSEQUENCES ON SENSE AMPLIFIERS

Process variations in deep sub-100-nm technology have become an ever-increasing concern, the effects of which are aggravated in deeply scaled transistors and at low V_{DD}. This causes a higher offset for the sense amplifier and results in a high failure rate or a low production yield in SRAMs. The CLSA shown in Figure 4.1 is a bilaterally symmetric circuit with the inverters of each branch (MP1–MN1 and MP2–MN2) cross coupled to form a latch. The input transistors (MN3 and MN4) receive input on their gate from the two bitlines. The data on the memory cell develops a lower voltage on one of the two bitlines compared to the other; thus, the two input transistors have different gate-to-source voltages. As a result, different currents flow through the two transistors, which results in slower discharge of one output node (OUT or OUT_B) compared to the other. Ultimately, the positive feedback mechanism provided by the cross-coupling the inverter's latches slows the discharging output node to V_{DD} and the other node to *GND*. The symmetry of the CLSA circuit is of critical importance for a reliable read operation in SRAMs, which employ differential sensing schemes. A mismatch in the corresponding devices of the two branches may neutralize or even reverse the effect of inputs and cause a read failure. For example, if the datum to be read is 1 but the V_{th} of MN3 is higher than the V_{th} of MN4, then even though the gate-to-source voltage of MN3 is higher than that of MN4, the current in MN4 might still be greater than MN3 due to an overall higher overdrive voltage. This will result in CLSA sensing the data as 0. Failing to address this problem can significantly degrade memory yield.

4.4.1 DIFFERENTIAL RESISTANCE MODELING OF CLSA

A metal oxide semiconductor field effect transistors (MOSFETs) can operate as a resistor whose value is controlled by the overdrive voltage ($V_{GS} - V_{th}$). Let us consider the operation of a CLSA in terms of the resistance of critical devices. Since the pull-down network contains the critical input transistors, much complexity can be avoided at negligible costs in terms of accuracy if we choose to analyze only the pull-down network resistance. During standby, all respective terminals of the corresponding transistors in the TRUE and COMPLEMENT branch have the same potential. Thus, $R_{MN1} = R_{MN2}$, $R_{MN3} = R_{MN4}$, and $\Delta R = 0$, as long as the dimensions and V_{th} of corresponding transistors are exactly matched. If the data on the memory cell is 1, then BL_B (Figure 4.1) begins to discharge, thus reducing the overdrive voltage of the input transistor (MN4) in the COMPLEMENT branch. R_{MN4} thus becomes greater than $R_{MN3,}$ and ΔR becomes negative. Subsequently, when the SEN transistor is switched ON this initial difference in the pull-down resistance in the two branches results in slower discharge of the complement output node as compared to that of the true output node. The positive feedback mechanism of the cross-coupled structure then ultimately latches to the slower discharging node (OUT_B) to V_{DD}, and the other output node is latched to *GND*. Similarly, opposite latching occurs when the datum on the memory cell is 0. In this way, the data on the memory cell are effectively translated to a differential pull-down resistance state in the sense amplifier. When the positive feedback mechanism is engaged, this differential pull-down resistance state resolves the output nodes to one of the two latched states depending on the polarity of ΔR (if ΔR is positive, OUT and OUT_B latch to 1 and 0, respectively, while if ΔR is negative, OUT and OUT_B latch to 0 and 1, respectively). Unlike conventional wisdom, we have investigated how the resistance of SEN devices plays a role in sense amplifier offset. If $V_{BL} = V_{BLB} = V_B$, then in symmetric CLSA, there will be *no* mismatch in the transistors. parameters of *MN1*, *MN2* and *MN3*, *MN4*. Here, the mismatches between *MP1* and *MP2* are unimportant because by the time these devices turn on the decision has been largely made by *MN1* and *MN2*. In this case, since $V_{BL} = V_{BL\_B} = V_B$ and the transistors are identical, we can write

$$V_{gs1} = V_{gs2}$$

Therefore, $I_{DSMN3} = I_{DSMN4}$ or we can write

$$I_{TRUE} = I_{COMPLEMENT}$$

From Figure 4.8, it is clear that

$$I_{DSN5} = I_{TRUE} + I_{COMPLEMENT}$$

Let us assume $I_{DSMN5} = I_{SS}$,

that is, for symmetric CLSA $I_{TRUE} = I_{COMPLEMENT} = I_{SS}/2$

Offset Correction in the Sense Amplifier

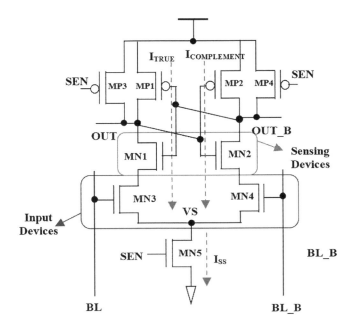

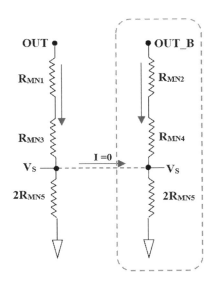

FIGURE 4.8 Half circuit resistance analogy of an asymmetric CLSA.

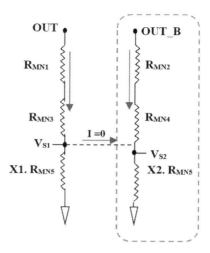

FIGURE 4.8 (Continued)

Then the circuit can be divided into two half circuits as shown in Figure 4.8b, and the differential resistance can be written as

$$\Delta_R = R_{TRUE} - R_{COMPLEMENT}$$
$$\Delta_R = (R_{MN1} - R_{MN2}) + (R_{MN3} - R_{MN4}) + (2R_{MN5} - 2R_{MN5})$$
$$\text{Here } R_{TRUE} = R_{MN1} + R_{MN3} + 2.R_{MN5}$$
$$R_{COMPLEMENT} = R_{MN2} + R_{MN4} + 2.R_{MN5}$$

and

$$\Delta_R = (R_{MN1} - R_{MN2}) + (R_{MN3} - R_{MN4})$$

for identical transistors, $R_{MN1} = R_{MN2}$ and $R_{MN3} = R_{MN4}$

Thus, $\Delta R = 0$

Offset Condition: Now consider if the V_{th} of MN3 is greater than the V_{th} of MN4. Let ΔV_{th} be equal to $(V_{th,MN3} - V_{th,MN4})$. In this case, even if both bitlines are at the same potential, the overdrive voltage of MN3 will be less than that of MN4, and hence, $R_{MN3} > R_{MN4}$. Now when BL_B discharges, R_{MN4} begins to increase as before, but it may or may not become greater than R_{MN3} before the positive feedback mechanism is engaged. In this way, the sense amplifier might latch to an undesired state causing a read failure,

that is, if $V_{TH,MN3} > V_{TH,MN4}$, than $I_{DSMN3} < I_{DSMN4}$, or we can write

$$I_{TRUE} < I_{COMPLEMENT} \qquad (3.1)$$

Offset Correction in the Sense Amplifier

Since, $I_{DSMNS} = I_{SS}$,

we can write $I_{TRUE} = \dfrac{I_{SS}}{X_1}$ and $I_{COMPLEMENT} = \dfrac{I_{SS}}{X_2}$.

From Equation 3.1, we can write $\dfrac{I_{SS}}{X_1} < \dfrac{I_{SS}}{X_2}$,

which employs $X_1 > X_2$.

Here, $\dfrac{X_1}{X_2}$ is the current ratio in the true and the complement branch.

From the schematic of a CLSA (Figure 4.8), it is clear that

$$I_{DSN5} = I_{TRUE} + I_{COMPLEMENT}$$
$$I_{SS} = \frac{I_{SS}}{X_1} + \frac{I_{SS}}{X_2}$$
$$\text{Thus,} \left(\frac{1}{X_1} + \frac{1}{X_2} = 1\right) \text{ or } X_1 = \frac{X_2}{X_2 - 1}$$

Therefore, with this analogy, the half circuit in the case of offset condition ($V_{TH,MN3} > V_{TH,MN4}$) is depicted in Figure 4.8. From Figure 4.8, the differential resistance is

$$\Delta R = R_{TRUE} - R_{COMPLEMENT}$$
Here $R_{TRUE} = R_{MN1} + R_{MN3} + X_1 \cdot R_{MN5}$
and $R_{COMPLEMENT} = R_{MN2} + R_{MN4} + X_2 \cdot R_{MN5}$
Thus, $\Delta R = (R_{MN1} - R_{MN2}) + (R_{MN3} - R_{MN4}) + (X_1 \cdot R_{MN5} - X_2 \cdot R_{MN5})$

It is important to note that since $X_1 \neq X_2$, this shows SEN device plays a role in differential resistance of CLSA in the case of offset ($V_{TH,MN3} > V_{TH,MN4}$). However, in the sense amplifier design, MN5 works as a current source, and by properly (up) sizing MN5, the R_{MN5} can be minimized, which makes the design less sensitive to R_{MN5} and it can be neglected.

Furthermore, it is important to note that until the time SA starts resolving the data, the node V_S will be shorted to *GND*, and R_{MN5} will be less significant.

4.4.2 Self-Correcting Sense Amplifier

This section explains the structure and functioning of the design that has been proposed in this chapter to minimize the input-referred offset. In all further discussions, the branch of the self-correcting sense amplifier (SCSA) that has the TRUE output node (OUT) will be denoted as the TRUE branch, and the one that has the complement output node (OUT_B) will be denoted as the COMPLEMENT branch.

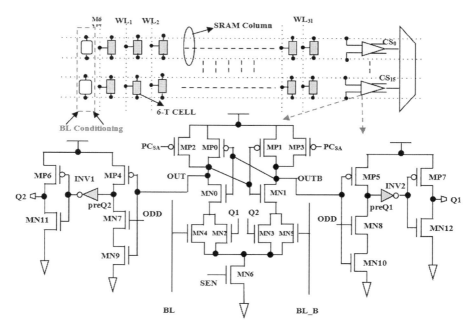

FIGURE 4.9 SRAM memory column with a detailed schematic of a self-correcting sense amplifier.

4.4.2.1 Design Overview

Figure 4.9 shows the schematic of the SCSA. The PMOS transistors (MP0, MP1) and NMOS transistors (MN0, MN1) constitute the two cross-coupled inverters. The nMOS transistors MN4 and MN5 are connected to true bitline (BL) and complement bitline (BL_B) at their respective gates and thus serve as input transistors in SCSA. MP2 and MP3 are the precharging devices to keep the output nodes at V_{DD} during the precharge phase. During precharge, the offset direction-determining (ODD) signal is kept high, which pulls the nodes preQ1 and preQ2 to ground (*GND*). Inverters INV1 and INV2 provide rail-to-rail voltage swings. Inverters MP6–MN11 and MP7–MN12 keep the final offset direction outputs—Q1 and Q2—at *GND* during standby.

4.4.2.2 Operation

Figure 4.10 shows the output, WL, and SEN signals during a read cycle. The read cycle proceeds in three phases. The first phase is the ODD phase. Figure 4.10 is reported for one read transaction for better understanding. Offset direction refers to which branch is more inclined to latch to V_{DD} compared to others in the absence of any bitline voltage difference. The SCSA stores this information on Q1 and Q2 nodes in the ODD phase. In the next phase, the differential voltage is developed on the bitlines. In the sensing phase, SCSA latches to one of the two states using the information from the first two phases regarding the data and the

Offset Correction in the Sense Amplifier

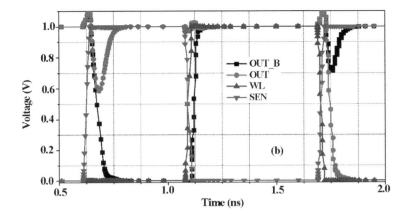

FIGURE 4.10 Simulated voltage characteristics at various nodes of SCSA in their respective phases.

offset direction. During the ODD phase, the bitlines remain precharged, and the SEN signal is asserted to allow the sense amplifier to latch. In the absence of any differential voltage on bitlines (ΔV_{BL}), the difference in currents through MN4 and MN5 comes only due to V_{th} mismatch. If V_{th}, MN4 > V_{th}, MN5 then I_{DS}, MN4 < I_{DS}, MN5, and thus, OUT is latched to V_{DD} while OUT_B is discharged to *GND* and vice versa. At the start of the ODD phase, the ODD signal is kept low. Depending on which of the two nodes (OUT or OUT_B) is pulled low, one of the two pMOS transistors, MP4 or MP5, is switched ON. This pulls preQ1 or preQ2 to V_{DD}, while the other preQ node will be close to *GND* (keeping its corresponding Q output tied to *GND*). In this way, if OUT is pulled low, Q2 rises to V_{DD} while Q1 remains at *GND*, and if OUT_B is pulled low, Q1 rises to V_{DD} while Q2 remains at *GND*. Thus, the information regarding the offset direction is captured on the Q nodes.

After the ODD phase, the differential voltage development phase begins as the WL is activated. During this phase, SEN is de-asserted, and the OUT and OUT_B nodes are precharged to V_{DD}. Note that since the ODD signal had been pulled low, the high preQ node is prevented from being pulled to *GND* despite the corresponding nMOS transistor (MN9 or MN10) switching ON as a result of the output nodes being precharged. During this time ΔV_{BL} is allowed to develop SCSA now has the information regarding its offset direction on the Q nodes and the information regarding data of the memory cell on the bitlines. Q1 and Q2 are connected to the gates of assist devices—MN2 and MN3 which are placed parallel to MN4 and MN5 respectively. If Q1 is high, the gate-to-source voltage of MN2 is higher than that of MN3 and vice versa. In this way, when the third phase begins with the assertion of the SEN signal, additional current flows in the TRUE branch if Q1 is high or in the COMPLEMENT branch if Q2 is high, compensating for the inherent V_{th} mismatch. The differential voltage on the bitlines can therefore now be sensed optimally and the sense amplifier latches to the desired state. A read access commences with the ODD phase by the activation of the SEN signal. This phase goes in parallel with the WL (*WL*) decoding time, so it incurs no timing overhead.

Here, it is important to note that in the case of small initial offset, that is, if the difference in voltage at bitlines, formed due to noise is sufficient to overcome intrinsic device offset, then a self-correcting loop may develop the opposite polarity of Q1 and Q2 and increase the offset in actual sensing phase. Thus, this approach is mainly intended to compensate for the effect of offset in sense amplifier devices and does not compensate for the noise effect. This is a topic for future research. Indeed, in SRAMs, bitlines in the cell array transition from a precharge state to a discharging state after activation of the WL. This current consumed in charging the bitlines leads to sensing noise. However, it is important to note that unique timing governs the three-phase operation in a read transaction of SCSA. A read access commences with the ODD phase by the activation of the SEN signal. This phase goes in parallel with the WL decoding time. Therefore, in the ODD phase, the *WL* is inactive; it reduces the probability of noise. Furthermore, the proper bitline conditioning circuit is used to minimize the small initial offset. For optimal offset reduction, the operating conditions (C_{BL}, ΔV_{BL}) must be noted, and accordingly, the strength of the assisting mechanism must be carefully adjusted to avoid incorrect latching caused by overcompensation. Mathematically, the optimal offset compensating should be just less than the differential voltage that develops on bitlines so that in the case of small favorable offsets, the compensation by assist devices does not reverse the effect of differential voltage on bitlines. It is because of such consideration that the assist devices are kept to a minimum size. Increasing the size of assistance would result in the SCSA being able to overcome greater unfavorable offsets but at the same time failing at favorable offset conditions. Since the calibration is applied to compensate for the offset that appears due to process variation, it is possible to perform the calibration process once at the start-up. However, to apply the calibration only at the start-up necessitates the unique timing and controlling of outputs (Q1 and Q2), since it is required when we need to apply the calibration and when we do not. This incurs an added cost in terms of additional clocked circuits like flip-flops with gates and control signals. To conduct calibration, once at start-up, the SCSA needs a minimum of two D flip-flops and one additional control signal as a flip-flop clock. Therefore, once at start-up, the CLK signal will transmit the information from OUT and OUTB to Q1 and Q2, and for the remaining cycle, it will run the SCSA in the conventional way with Q1 and Q2. Moreover, it increases the area overhead due to a minimum of 20 more transistors in calibration circuitry. This results in more power consumption in SCSA.

4.4.2.3 Offset Analysis in SCSA

As shown in Figure 4.9 the pull-down network in SCSA is modified from that of a CLSA by adding an additional transistor, parallel to each input transistor. This results in $R_{TRUE} = R_{MN0} + (R_{MN4}||R_{MN2}) + X1.R_{MN6}$ and $R_{COMPLEMENT} = R_{MN1} + (R_{MN5}||R_{MN3}) + X2.R_{MN6}$. The presence of MN2 and MN3 provides a calibration scope to this design. So, for example, if $\Delta V_{th} > 0$, then $R_{MN4} > R_{MN5}$. To oppose this offset in ΔR, R_{MN3} is made greater than R_{MN2} by controlling the overdrive voltage of the assist devices

Offset Correction in the Sense Amplifier 141

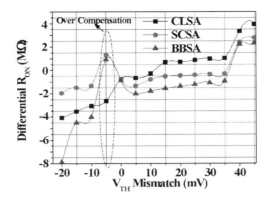

FIGURE 4.11 Differential resistance variation with changes in ΔV_{th}.

(MN2 and MN3) using a calibration circuitry as was explained in Section 4.3.1. Thus, it is important to note that in the SCSA, due to the calibration mechanism, even in the offset condition $(V_{THN3} > V_{THN4})$, the $I_{TRUE} > I_{COMPLEMENT}$, that is, $X1 < X2$, it causes the term $(X1.R_{MN6} - X2.R_{MN6})$ to become negative and contribute to maintaining the more positive R. Since R_{MN6} is the same in CLSA and SCSA, the term $(X1.R_{MN5} - X2.R_{MN5})$ will be common in the differential resistance analogy of CLSA and SCSA; thus, it will not have the significance in comparative analysis of CLSA and SCSA.

In body bias sense amplifier (BBSA), the input transistors are in the pull-up network. It can be seen in Figure 4.7 that for BBSA, $R_{TRUE} = R_{MP1} + R_{MP3} + 2.R_{MP5}$, $R_{COMPLEMENT} = R_{MP2} + R_{MP4} + 2. R_{MP5}$ and $\Delta R = (R_{MP1} - R_{MP2}) + (R_{MP3} - R_{MP4})$. Let us define ΔV_{th} for BBSA as $(|V_{th, MP3}| - |V_{th,MP4}|)$, when the bodies of both input transistors are biased to the same voltage. Here, we define ΔV_{th} for BBSA as $(|V_{th, MP3}| - |V_{th,MP4}|)$, when the bodies of both input transistors are biased to the same voltage. Figure 4.11 manifested the differential R_{ON} vs. V_{th} mismatch curve by introducing the V_{th} mismatch, which illustrates the significant finding of the calibration schemes (SCSA and BBSA) in terms of sharp switching at 0 crossing. Here, it is important to note that for a small favorable offset $(V_{th, MN4} < V_{th, MN5})$, due to the digital nature of the calibration mechanism, the assistance strength is higher than the offset, which further increases the offset (+ve differential R_{ON}). However, this issue can be resolved by properly selecting the $Q1/Q2$ voltage, less than the V_{DD} (which requires a dual supply).

Furthermore, to account for the worst-case mismatch condition ΔR is plotted against ΔV_{th} in Figure 4.12 for CLSA, SCSA, and BBSA by a 2,000-point Monte Carlo simulation. Data in all three cases are set such that for $\Delta V_{th} = 0$, ΔR in all cases is negative (read '1' case). This graph illustrates how CLSA, which does not have an offset compensating mechanism, suffers from a much steeper change in ΔR as ΔV_{th} deviates from its ideal value of 0, compared to SCSA and BBSA. The ΔR value, which should remain negative for reliable sensing, becomes positive in CLSA at a lower value of ΔV_{th} compared to that in SCSA and BBSA. This characteristic makes SCSA and BBSA more robust and suitable for high-reliability applications.

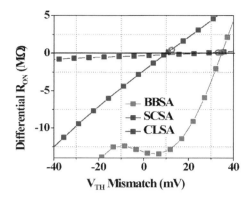

FIGURE 4.12 Variation in differential resistance in the TRUE and COMPLEMENT branch with changes in ΔV_{th}.

For a fair comparison, the device parameters are kept at equal values, that is, sensing devices, input devices, and SEN devices are 1.2μm for CLSA, SCSA and BBSA. To save the area, the calibrating devices MN2–MN3 are made using minimum-sized devices. The sizing of MN2 and MN3 significantly affects the resistance of each pull-down path. For larger device sizes, the effective resistance of the pull-down path will reduce significantly because the resistance term $(R_{MN4}||R_{MN2})$ and $(R_{MN5}||R_{MN3})$ will reduce. However, for larger device sizes, the resistance term $(R_{MN4}||R_{MN2})$ and $(R_{MN5}||R_{MN3})$ will reduce. However, for larger device sizes, the capacitance introduced by these devices tends to slow down the process of sensing. Furthermore, in view of the area, large-sized calibrating devices are also not preferable.

To maintain this area, for delay trade-off, we kept the optimal size of MN2 and MN3. It is important to note that the calibration device sizing is not very significant in the ODD phase since, in the ODD phase, Q1 and Q2 are at logic zero. Therefore, MN2 and MN5 (Figure 4.9) will be in the cut-off range and will not contribute to determining the offset direction. In the actual sensing phase, the overdrive voltage of MN2 and MN5 will be mostly governed by their gate voltage (Q1 and Q2), making it less sensitive to their V_{th} offset.

4.5 DISCUSSION OF IMPORTANT FIGURES OF MERITS

CLSA, SCSA, and BBSA were simulated using the standard 65nm bulk CMOS process, and their input-referred offset was analyzed along with several other performance parameters like sensing delay and power consumption. A standard 6T SRAM was used as the memory cell for the SRAM array with all three sense amplifiers. Since BBSA uses PMOS devices as input transistors, bitlines, in the case of BBSA, are initially precharged to *GND*, and one of the bitlines (depending on the data stored) begins to charge to positive voltage via the pull-up network of the 6T cell. This scheme is different from the case of CLSA and SCSA as they use NMOS

Offset Correction in the Sense Amplifier   143

devices as input transistors, and therefore, the bit-lines are charged to V_{DD} initially and one of the bitlines discharges via the pull-down network of the 6T cell. To normalize this difference and produce fair comparative results, the pull-down strength in the case of CLSA and SCSA is matched to the pull-up strength of BBSA so that the same differential voltage is developed on the bitlines in the same amount of time. All other corresponding transistors in CLSA, SCSA, and BBSA have the same sizes for a fair comparison.

The calibrating circuitry in SCSA and BBSA is implemented using minimum-sized devices. Simulations were carried out at various PVT corners as pointed out in the text. However, the default operating conditions were set as follows: Data on the memory cell was 1. Bitlines were loaded with an external capacitance of 500fF each to gauge the effect of the large SRAM macro. The operating temperature was set to 27°C, and simulations were carried out at the TT process corner. The primary supply voltage was set to 700mV. In the case of BBSA, a second voltage value is applied to the body of one of the input transistors (as decided by the calibration circuitry).

4.5.1 OFFSET MEASUREMENT

This section analyses the distribution of input-referred offset in a 2000-point Monte Carlo simulation carried out on CLSA, SCSA, and BBSA. It further explores how the currents in the two branches of sense amplifiers affect the offset distribution. For the validation of the proposed sense amplifier, all the sense amplifiers in being compared, that is, CLSA, BBSA, and the proposed circuit, have been implemented with 4-kb SRAM array in a 65nm standard CMOS process. The array comprised 8 blocks, each comprising 16 columns of 32 6T SRAM cells. Each column had one sense amplifier, and thus, the 4-kb SRAM array comprised 128 sense amplifiers. Each column of the bank has one local sense amplifier. The orders in which the memory cells are activated are identical for all four designs. During read, the unselected banks are kept in standby mode, and only one bank will be active. For the measurement, each column contains 32-bit cells. Furthermore, an additional 500fF capacitance is connected to the bitline to model additional parasitic capacitance in the bigger SRAM macro.

As mentioned in Section 4.3.1, with an increase in V_{th} mismatch, the number of currents flowing in the TRUE and COMPLEMENT branches (during sensing) changes. For data 1, all three sense amplifiers require a higher current to flow in the TRUE branch, and for data 0, a higher current needs to flow to the COMPLEMENT branch. Assuming data 1, as ΔV_{th} increases, the current in the TRUE branch begins to decrease and, in the COMPLEMENT branch, begins to increase. The average drain to source current flowing in the TRUE and COMPLEMENT branches' transistors on the path from output nodes to the respective rails was measured for 2,000 Monte Carlo iterations in all three sense amplifiers.

The scatterplot of TRUE current and COMPLEMENT current against V_{th} difference between sensing transistor in the TRUE branch and that in the COMPLEMENT branch (ΔV_{th}) is plotted in the form of a fourth-order polynomial approximation for better accuracy in Figure 4.13. It can be observed that in the

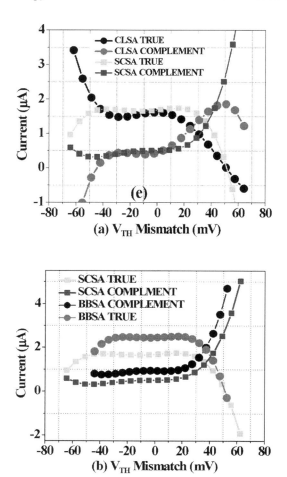

FIGURE 4.13 I_{TRUE} and $I_{COMPLEMENT}$ versus ΔV_{th} during the sensing phase (a) CLSA and SCSA and (b) BBSA and SCSA.

case of SCSA and BBSA, the current in the TRUE branch remains higher than that in the COMPLEMENT branch for a larger ΔV_{th}, which is illustrative of the fact that SCSA and BBSA can tolerate higher V_{th} mismatch in their sensing transistors. BBSA and SCSA both maintain higher current in the TRUE branch for ΔV_{th} lower than about 37mV while CLSA does so for ΔV_{th} lower than about 25mV. The higher current in the TRUE branch of SCSA is due to the reduction of equivalent resistance from OUT to *GND* by activating the assist device MN9, which is parallel to the input transistor. While in BBSA higher current in the TRUE branch is due to the reduction of equivalent resistance from OUT to V_{DD} by assigning lower body voltage to the input transistor in the TRUE branch. Ultimately, this differential current, which maintains its polarity for a higher range of V_{th} mismatch in the case

Offset Correction in the Sense Amplifier

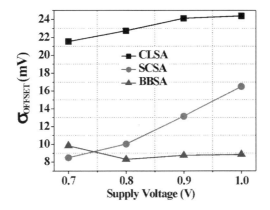

FIGURE 4.14 Standard deviation of input-referred offset for the designs at different V_{DD}.

of BBSA and SCSA as compared to CLSA, causes a reduction in the input-referred offset of BBSA and SCSA.

Figure 4.14 shows that the standard deviation of offset distribution in SCSA design becomes greater than that of BBSA at higher supply voltages. This happens because with an increase in primary supply voltage, the secondary voltage of BBSA is also increased, and so the compensation provided remains close to optimal. While in SCSA, the width of the assist devices (MN2 and MN3) has been optimized for 700mV. This happens because with an increase in primary supply voltage, the secondary voltage of BBSA is also increased, and so the compensation provided remains close to optimal. While in SCSA, the width of the assist devices (MN2 and MN3) has been optimized for 700mV operation, and with an increase in supply voltage, this width is not calibrated. Thus, for higher supply voltages, SCSA tends to overcompensate (reverse favorable ΔR polarity) the V_{th} mismatch of sensing transistors, thus widening its input-referred offset distribution. Therefore, it is essential in the case of SCSA to optimize the width of assist devices according to the supply voltage. Nevertheless, SCSA's offset distribution has a lower standard deviation than that of conventional CLSA even without any width optimization.

4.5.2 Performance Analysis

CLSA, SCSA, and BBSA were compared on important performance metrics such as active power consumption and sensing delay. All three designs were tested at various process, voltage, and temperature corners. Sensing delay for a sense amplifier is the time it takes for the dropping or rising (depending on whether the output nodes are precharged to V_{DD} or GND) output to reach 50% of the full swing after the SEN signal has reached 50% of its full swing potential. BBSA requires separate well potentials for the two input transistors, and thus, PMOSs have to be used for the input transistors. The rising output node is charged via the pull-up network of PMOS devices in BBSA, while in the case of SCSA and CLSA, the falling output is discharged via the pull-down network of NMOS devices. Since the mobility of electrons is higher than that of holes, PMOS devices of the same dimensions are inherently slower than NMOS devices. Therefore, CLSA and SCSA have a much smaller sensing delay than BBSA. At nominal conditions, the sensing delay in SCSA and CLSA is about 78% and 77%, respectively.

At nominal conditions, the sensing delay in SCSA and CLSA is about 78% and 77% lower than that in BBSA, respectively. We can see in Figure 4.15 that SCSA and CLSA have a lower sensing delay than BBSA for all process corners. The sensing delay in SCSA is 63% to 87% less, compared to BBSA. As was discussed earlier, SCSA has two additional assist devices, placed parallel to the input transistors. These devices result in two opposing tendencies as far as sensing delay is concerned. The capacitance introduced by these devices tends to slow down the process of sensing, while the additional current path provided by these devices tends to increase the speed of sensing. Thus, the sensing delay is competitive in the two designs (CLSA and SCSA) in various analyses.

In the case of the offset in input devices (MN4 and MN5) the Q1 or Q2 will be at either logic 0 or 1 after the ODD phase, resulting in one of the devices MN2 or MN3 being turned on and will provide the additional current or improved strength. However, in the case of the no-offset condition, since Q1 and Q2 will be at logic 0, MN2 and MN3 will be cut off, no additional path for the current will be there, and SCSA possesses the almost-equal delay than CLSA. Figure 4.15 reports the sensing delay in no offset condition. This shows SCSA has competitive sensing delay with no delay overhead. Figure 4.16 shows the sensing delay distribution of CLSA, SCSA, and BBSA obtained from 1,000-iteration Monte Carlo simulations at 0.7V supply. It can be seen that all BBSA instances have a much greater sensing delay than CLSA and SCSA. Also, while there is a very small (0.7%) increase in mean sensing delay in SCSA as compared to CLSA, regarding variability, SCSA is better with about 31% lower standard deviation. Results at other supply voltages have been summarized

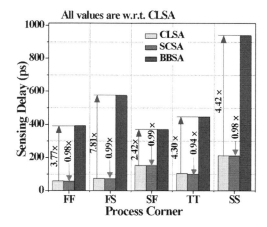

FIGURE 4.15 Sensing delay in CLSA, BBSA, and SCSA at different process corners.

Offset Correction in the Sense Amplifier 147

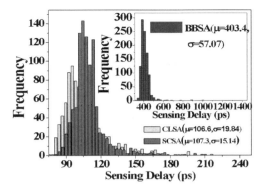

FIGURE 4.16 Distribution of sensing delay in CLSA, BBSA, and SCSA.

TABLE 4.1
Average Sensing Delay and Standard Deviation in CLSA, SCSA, and BBSA for Different Supply Voltages

V_{DD}	μ_{CLSA} (ps)	σ_{CLSA} (ps)	μ_{SCSA} (ps)	σ_{SCSA} (ps)	μ_{BBSA} (ps)	σ_{BBSA} (ps)
0.7 V	106.6	19.84	107.3	15.14	403.4	57.07
0.8 V	69.43	9.334	70.16	6.557	246.4	20.65
0.9 V	50.63	5.452	51.42	3.535	171.8	16.09
1.0 V	39.31	2.85	40.42	2.312	127.2	9.294

in Table 4.1. These results suggest that while the capacitive effects tend to increase delay, in SCSA, CLSA is favored in some cases; in most cases, SCSA stands superior to CLSA as far as sensing delay is concerned. BBSA, by comparison, is much slower than both CLSA and SCSA.

Active power consumption has been measured as the average power consumed during a complete read cycle in all three designs. The prediction phases have been included for both BBSA and SCSA. Both BBSA and SCSA have calibrating circuitry that has several devices. This results in more capacitance switching than that in CLSA. Therefore, power consumption is higher in BBSA and SCSA as compared to CLSA. Since SCSA has fewer devices than BBSA and thus has lower power consumption as well, at a nominal corner (0.7V supply, TT process, 27°C temperature), the power consumption in SCSA is about 25% more than that in CLSA, and BBSA consumes about 56% more power than CLSA. Figure 4.17 illustrates that this trend is followed across all process corners. The power consumption in SCSA is 14% to 45% more than that in CLSA, and in the case of BBSA, it is about 42% to 83% more than CLSA. These results suggest that while both BBSA and SCSA provide higher offset compensation than CLSA at significant costs in terms of power consumption,

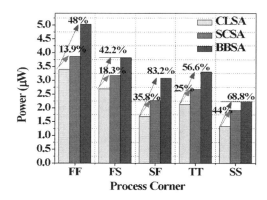

FIGURE 4.17 Active power consumption in CLSA, BBSA, and SCSA at different process corners.

SCSA consumes much less power than BBSA. However, as the input-referred offset reduces, the required differential voltage on bitlines to provide 100% yield also reduces. Since the voltage swing is reduced on high capacitance bitlines for low-offset sense amplifiers, the memory consumes lower power than in the case of CLSA.

Figure 4.18 shows the active memory power consumption in memory and the sense amplifier when the differential voltage on the bitlines in each case is adjusted to achieve a 100% yield. Interestingly, memory power consumption in the case of BBSA and SCSA is 34% and 27% lower than that in the case of CLSA, respectively. Figure 4.19 presents an illustrative summary of the performance characteristics of the three SAs in terms of their power delay products at different process corners. It can be observed that SCSA has only slightly higher power delay products than CLSA (0.12× to 0.19× higher), while BBSA has significantly higher power delay products than SCSA (8.5× higher at the FS corner). Figure 4.20 shows the yield of CLSA, SCSA, and BBSA at various differential voltages on bitlines.

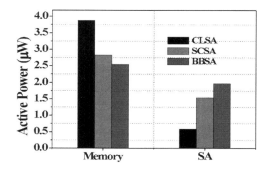

FIGURE 4.18 Power consumption in memory and the sense amplifier when operation carried out for 6σ yield.

Offset Correction in the Sense Amplifier 149

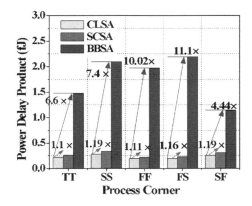

FIGURE 4.19 SA power delay product of CLSA, SCSA, and BBSA.

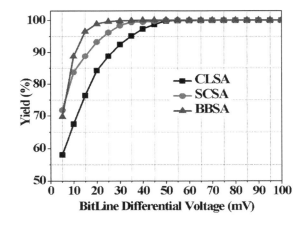

FIGURE 4.20 Sense amplifier yield for different values of differential bitline voltages.

Although SCSA consumes more power than CLSA, the yield of SCSA topology is much greater than that of CLSA and sensing delay is also generally lower in the case of SCSA. Table 4.2 shows performance summaries of CLSA, SCSA, and BBSA when employed in the 4-kb array. At the array level, worst-case analysis was carried out for sensing delay (at the SS process corner and −40°C) and active power consumption (at FF process corner and 125°C). The results correspond to the results obtained in cell-level analysis following the same trend. While both SCSA and BBSA consumed more power than CLSA due to additional devices in a calibrating circuit, BBSA consumed from 1.9% more to 25.7% more power than SCSA at various supply voltages.

In the case of sensing delay, CLSA and SCSA had comparable values as can be seen in Figure 4.21. BBSA, however, had a sensing delay 1.6× to 2.9× the sensing

TABLE 4.2
Performance Summary of CLSA, SCSA, and BBSA When Employed in 4-kb SRAM Array

SA design	Sensing delay (ps)	Active power (µW)	Power delay products (fJ)
CLSA	1331	1.62	2.156
SCSA	1324	3.3	4.369
BBSA	2130	4.15	8.839

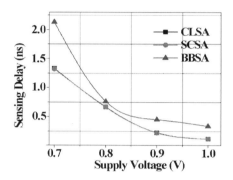
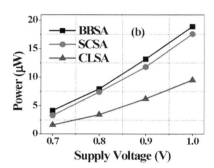

FIGURE 4.21 Worst-case (a) sensing delay and (b) active power in sense amplifiers when employed in implemented 4-kb SRAM array.

delay in SCSA. These results further confirm the superior performance of SCSA compared to BBSA. Furthermore, it is important to note that SCSA has no delay overhead in comparison to CLSA.

4.6 AREA CONSIDERATION

Custom layouts of the SCSA have been designed in the commercial 65nm standard CMOS process and the array has been generated using instances. For area comparison, the layouts of other designs are also implemented with full adherence to the same design rules for a fair comparison, rather than comparing the area to previously reported 'pushed rules' implementations. SCSA and other compared sense amplifier designs follow the same bitline pitch. The resulting layout shown in Figure 4.22 fits into the 22.25µm² footprint, which is approximately 1.36× larger than CLSA; however, SCSA possesses 0.31· (68%) less area than BBSA (69.66µm²). Since we were tackling the challenge posed due to a mismatch in corresponding transistors, our design strategy was to make the circuit as symmetric as possible. Each layer-to-layer contact has been provided with a redundant contact to increase the robustness of the structure. To avoid performance degradation, long poly routing has been avoided. To save area, the calibrating circuit (MN7–MN12) and (MP4–MP7) is made using

Offset Correction in the Sense Amplifier

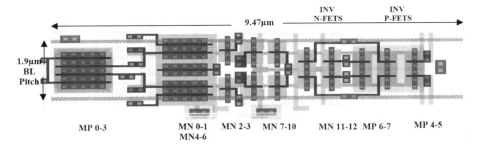

FIGURE 4.22 Layout of SCSA.

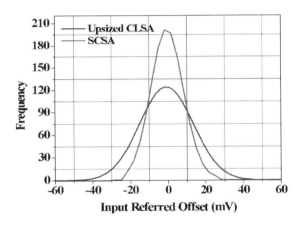

FIGURE 4.23 Input-referred offset distribution in upsized CLSA and SCSA.

minimum-sized devices. Even though SCSA occupies 1.36× more area than CLSA, its better built-in process tolerance the overwhelming improvement in terms of input-referred offset justifies the area overhead. Furthermore, this area overhead is amortized by the height of the memory cell array (i.e., number of cells on a column). Furthermore, for a fair comparison with the SCSA, post-layout circuit simulations were performed for the iso-area conditions (upsized input devices in CLSA). Compared to upsized CLSA, SCSA exhibits a tighter offset distribution area as can be manifested in Figure 4.23. The physical layout of the sophisticated 4-kb SRAM array in 65nm standard CMOS technology is shown in Figure 4.24.

4.7 CHAPTER SUMMARY

This chapter presented a novel sense amplifier topology and compared the proposed design to conventional CLSA and another offset-compensating sense amplifier—BBSA. The operating mechanism and offset problem were discussed in terms of the resistance of critical devices in the sense amplifier, explaining how the data on a memory cell is effectively translated to a differential resistance polarity in the sense amplifier and how V_{th} mismatch can result in incorrect differential resistance polarity in the sense amplifier. Reliability in terms of input-referred offset of the three sense amplifiers was discussed along with other performance parameters, such as sensing delay and power.

REFERENCE

1. P. Singh et al., "Device/Circuit/Architecture techniques for ultra-low power FPGA design," *Journal of Solid State Electronics*, vol. 2, no. 1, pp. 1–15, 2013.
2. E. Seevinck, P. J. van Beers and H. Ontrop, "Current-mode techniques for high-speed VLSI circuits with application to current sense amplifier for CMOS SRAM's," *IEEE Journal of Solid-State Circuits*, vol. 26, no. 4, pp. 525–536, April 1991, doi:10.1109/4.75050.
3. B. Wicht, T. Nirschl and D. Schmitt-Landsiedel, "Yield and speed optimization of a latch-type voltage sense amplifier," *IEEE Journal of Solid-State Circuits*, vol. 39, no. 7, pp. 1148–1158, July 2004, doi:10.1109/JSSC.2004.829399.
4. D. Anh-Tuan, K. Zhi-Hui and Y. Kiat-Seng, "Hybrid-mode SRAM sense amplifiers: New approach on transistor sizing," *IEEE Transactions on Circuits and Systems II: Express Briefs*, vol. 55, no. 10, pp. 986–990, October 2008, doi:10.1109/TCSII.2008.2001965.
5. M. J. M. Pelgrom, A. C. J. Duinmaijer and A. P. G. Welbers, "Matching properties of MOS transistors," *IEEE Journal of Solid-State Circuits*, vol. 24, no. 5, pp. 1433–1439, October 1989, doi:10.1109/JSSC.1989.572629.
6. S. Borkar, T. Karnik, S. Narendra, J. Tschanz, A. Keshavarzi and V. De, "Parameter variations and impact on circuits and microarchitecture," *Proceedings 2003. Design Automation Conference (IEEE Cat. No.03CH37451)*, 2003, pp. 338–342, doi:10.1145/775832.775920.
7. S. J. Lovett et al., "Yield and matching implications for static RAM memory array sense amplifier design," *IEEE Journal of Solid-State Circuits*, vol. 35, no. 8, pp. 1200–1204, 2000.
8. B. S. Reniwal, P. Bhatia and S. K. Vishvakarma, "Design and investigation of variability aware sense amplifier for low power, high speed SRAM," *Microelectronics Journal, Elsevier*, vol. 59, pp. 22–32, January 2017.

5 Data Sensing in SRAM: A Hybrid Approach with FinFET

5.1 INTRODUCTION

Voltage scaling is widely adopted to improve energy efficiency in modern SRAMs. As the V_{DD} is scaled down, the decreasing memory cell current (I_{CELL}) issue, concerning differential current degradation in the current sense amplifier, becomes severe. Therefore, SRAM suffers from a decrease in read cell current as the device size and V_{DD} shrink down, while the threshold voltage remains the same. The vulnerability to decreased I_{CELL} is exacerbated in the following situations. With continued efforts to reduce the size of the SRAM cell and the bitline (BL) pitch, overcoming these issues has become a major challenge in the read operation of SRAMs. The SRAM voltage/technology scaling challenge is illustrated in Figure. 5.1, which plots the reported 6T SRAM die-level or macro-level current against the technology node for the past years. This affirms that I_{CELL} is constantly reducing with scaling.

1. Small cell area in order to achieve a smaller area per bit
2. Low V_{DD} [1–4] for reduced power consumption

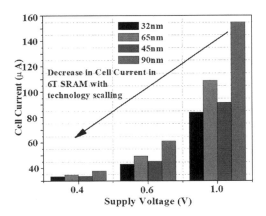

FIGURE 5.1 The SRAM technology scaling barrier: Reported 6T SRAM versus technology node. Array sizes range from 64 kb to 512 Mb.
Source: ISSCC and VLSI (2007–2018).

DOI: 10.1201/9781003213451-5

5.2 DESIGN BOTTLENECKS IN CURRENT MODE SENSING IN A SCALED COMPLEMENTARY METAL-OXIDE SEMICONDUCTOR PROCESS

For data sensing many small memories (nonvolatile memories and low-voltage SRAMs) employ a voltage mode sense amplifier (VSA) [5–12] with long *BL* developing time to provide tolerance against *BL* and SA offset. However, this is accomplished at the cost of a reduction in read speed. Over the last two decades, a number of current mode sense amplifiers (SAs) have been proposed, aiming to improve sensing speed and power consumption during the read operation in SRAM [5,7,10,13–19]. Since all these SA designs utilize the differential output current of the current conveyor, their improvement is only incremental. Seevinck et al. [20] clearly indicate that the differential current is equal to the current flowing into the cell node with a "0" stored in it, that is, I_{CELL}. Our analysis and simulations have proved otherwise. In fact, the differential current is much smaller than the I_{CELL}, due to imperfection in the current conveyor. In all current sense amplifiers (CSAs) and hybrid sense amplifiers (HSAs), the differential current develops the differential voltage on datalines (DLs). However, with complementary metal-oxide semiconductor (CMOS) technology scaling, the dataline capacitance may become too large. This results in insufficient differential voltage to be developed by the differential current. To overcome the mismatch issue, the differential voltage on the DLs should at least reach up to the SA offset voltage. Thus, offset in combination with huge DL capacitance results in small I_{CELL} SRAMs suffering from slow read speed and high read fail probability. Therefore, developing an SA with greater offset tolerance is essential to achieve high yield in small I_{CELL} SRAMs with fast read operations.

Despite such research activity in SA design, there are very few implementations of SAs in fin-shaped field-effect transistors (FinFETs) or multigate FETs (MuGFETs) [3,21–25]. Furthermore, a comprehensive assessment of FinFET SAs, considering local random variation in SA devices and variation of SA offset voltage (for differential sensing), is still lacking. In this work, for the first time, a dataline-isolated FinFET current mode SA, targeted to small I_{CELL} sensing in SRAM, is presented with a detailed sensitivity analysis. To the best of our knowledge, this is the first work that evaluates read acceleration through boosting differential current in FinFET current mode SA for SRAM. Besides reducing the sensing delay, it also improves read yield and decreases the effect of dataline capacitance of SRAMs. Furthermore, the current feed (CF)-SA is able to tolerate a large offset of the input devices, up to 80mV.

5.3 DIFFERENTIAL CURRENT MODE SA DESIGN

The conventional current mode SA (CCSA) consists of four equally sized p-type metal-oxide semiconductor (PMOS) transistors as shown in Figure 5.2. In most cases, it can fit within the column pitch, avoiding the need for column-select devices, thus further reducing propagation delay. In such architectures, the column decoder output is connected to the CS signal, which particularly selects the column. Therefore, decoding occurs through P2 and P3 devices. However, this approach is limited to current mode sense amplifiers only. Transistors P4 and P5 are the precharge transistors, and P6 is the equalization transistor.

Data Sensing in SRAM: A Hybrid Approach with FinFET

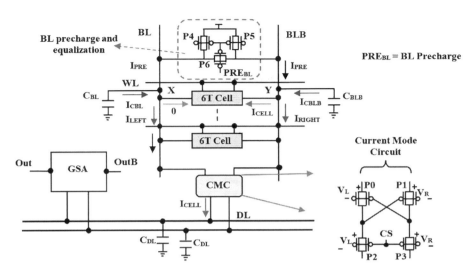

FIGURE 5.2 Detailed schematic of CCSA with memory column in typical SRAM architecture.

PRE_{BL} is the bitline precharge signal. Before sensing, datalines are precharged to ground. The circuit operates as follows. Consider that a particular cell is accessed and that it draws current I_{CELL}. In our analysis, I_{CELL} is defined as the instantaneous (maximum) current during the WL activation window. Now, once the column select (CS) signal goes low, the gate–source voltages of P0 and P2 become equal, since their currents and sizes are equal and both transistors are in saturation. This voltage is denoted by V_L. The same applies to P1 and P3. Their gate–source voltages are denoted by V_R. It follows that since the CS signal is grounded and the datalines are precharged to ground, both the left and right bitlines will have voltage $V_L + V_R$. Thus, the potential of the bitlines will be equal independent of the current distribution. This means that there exists a virtual short circuit across the bitlines. Since the bitline voltages are equal, the bitline load currents will be also equal, and the bitline capacitor currents will be also equal. As the cell draws current I_{CELL}, it follows that the left leg of the SA must pass more current than the right leg. In fact, the difference in these currents is I_{CELL}, the cell current. The drain currents of P2 and P3 are passed to the current-transporting datalines DL and DLB. The datalines' differential current is thus equal to the cell current. The differential current analysis of current mode circuit (CMC) is as follows:

$I_{LEFT} = I_{PRE} + I_{CBL}$

$I_{PRE} + I_{CBLB} = I_{CELL} + I_{RIGHT}$

That is, $I_{RIGHT} = I_{PRE} + I_{CBLB} - I_{CELL}$

Thus, the differential current (I_{DIFF}) is $I_{LEFT} - I_{RIGHT} = I_{PRE} + I_{CBL} -$
$$(I_{PRE} + I_{CBLB} - I_{CELL})$$

That is, $I_{DIFF} = I_{CELL}$.

Therefore, the differential current is equal to the cell current. In this way, we obtain current sensing. However, with CMOS technology scaling, the dataline capacitance (C_{DL}, C_{DLB}) may become too large, and the voltage difference produced by the differential current which is equal to the I_{CELL} may be insufficient to meet the offset tolerance requirement of global sense amplifier (GSA). Therefore, CCSA must develop a first-stage voltage difference on the heavily loaded datalines using continuous I_{CELL} driving. Thus, the read access time of the CCSA is sensitive to DL load and mismatch.

5.4 DIFFERENTIAL CURRENT FEED TECHNIQUES

Figure 5.3 shows a simple schematic illustration of the proposed CF-SA with SRAM column and is followed by the CF-SA timing diagram in Figure 5.4. The CF-SA uses an isolated-dataline approach to rapidly generate the first-stage voltage difference on small-load internal nodes (X1 and X2). This is possible due to its decoupled DLs. In addition, the feed currents (I_L, I_R) are not sensitive to a mismatch in the transistors and C_{DL}. Due to C_{DL}/C_{DLB} decoupling, the differential potential on the datalines (V_{DIFF}) increases, resulting in reduced sensing delay. The proposed scheme isolates the column-select devices (P2 and P3) from the huge dataline capacitance. This is accomplished by adding transistors N1 and N2. Thus, nodes X1 and X2 are no longer loaded with the huge dataline capacitance and instead have small device capacitance of P2–P3, N1–N2, and inverter (INVL/INVR) devices. It is worth noting that two p-type field effect transistors (PFETs)—P4 and P5—are used to drive the data-line capacitance based on the input voltage.

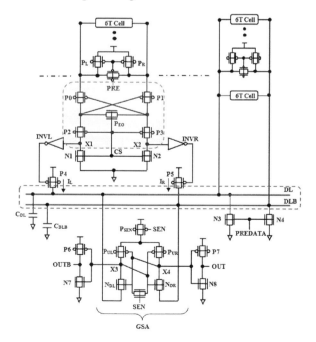

FIGURE 5.3 Schematic of current feed SA with a memory column.

Data Sensing in SRAM: A Hybrid Approach with FinFET   157

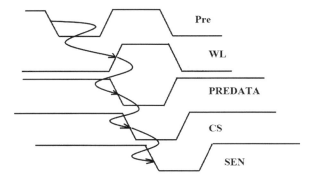

FIGURE 5.4 Timing diagram of current feed SA.

5.5 CAPACITANCE MODELING OF SENSING CIRCUIT

Figure 5.5 shows the approximate capacitive model of FinFET of CCSA. Once the WL is activated, the node storing "0" will have the current I_{CELL}. As the cell draws current I_{CELL}, it follows that the left-hand leg of the SA must pass more current than the right-hand leg. The drain currents of P2 and P3 are passed to current-transporting datalines DL and DLB. The datalines' differential current is therefore equal to the cell current. Thus, we obtain current sensing. The differential current develops the voltage at the datalines. Hence, the differential voltage at DL, V_{DIFF}, can be characterized as

$$\Delta V_{Diff} = |V_{DL} - V_{DLB}|$$
$$= \left| \int_0^{T_{ACC}} \frac{1}{C_{X1}} I_L \cdot dt - \int_0^{T_{ACC}} \frac{1}{C_{X2}} I_R \cdot dt \right|. \quad (5.1)$$

Here, T_{ACC} is the access time:

$$\text{For } C_{X1} = C_{X2} = C_X$$
$$\Delta V_{Diff} = \left| \frac{1}{C_X} \int_0^{T_{ACC}} (I_L - I_R) \cdot dt \right|$$

Here $I_L = I_{BL-PRE} - I_{C_{BL}}$
$I_R = I_{BL-PRE} - I_{CELL} - I_{C_{BLB}}$.

Since the actual difference current $|I_L - I_R|$ in the case of CCSA is the cell current I_{CELL},

$$\Delta V_{DIFF} = \left| \frac{1}{C_X} \int_0^{T_{ACC}} I_{CELL} \cdot dt \right|. \quad (5.2)$$

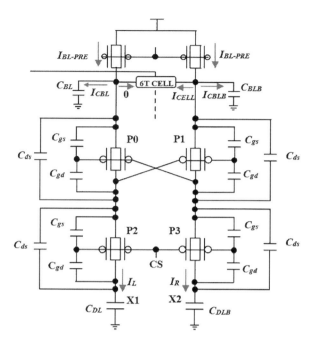

FIGURE 5.5 Capacitive model of current mode circuit in a CCSA.

In the case of CCSA, the capacitance at node X1 is

$$C_{X1} = (C_{ds_{fr}} + C_{gd})_{P2} + C_{DL}. \tag{5.3}$$

Here $C_{ds_{fr}}$ = Drain to source fringing capacitance,

$$C_{gd} = \text{gate to drain capacitance}$$
$$C_{gd} = C_{gd_{OV}} + C_{gd_{fr}},$$

$C_{gd_{OV}}$ = gate to drain overlap capacitance, and
$C_{gd_{fr}}$ = gate to drain fringing capacitance.

The effective capacitance at node X1 is

$$C_{X1} = (C_{ds_{fr}} + C_{gd_{OV}} + C_{gd_{fr}}) + C_{DL}$$
$$\text{Here} \quad (C_{ds_{fr}} + C_{gd_{OV}} + C_{gd_{fr}}) \ll C_{DL}. \tag{5.4}$$

Similarly, the capacitance at node X2 is

$$C_{X2} = (C_{ds_{fr}} + C_{gd_{OV}} + C_{gd_{fr}})_{P3} + C_{DLB} \tag{5.5}$$

Data Sensing in SRAM: A Hybrid Approach with FinFET

Figure 5.6 shows the capacitive modal of CFSA. In Figure 5.6, it is shown that the capacitance at node X1 is

$$C_{X1} = (C_{ds_{fr}} + C_{gd})_{P2} + (C_{ds_{fr}} + C_{gd})_{N1} + C_{gsPINVL} + C_{gsNINVL}$$

Here $C_{gd} = C_{gd_{OV}} + C_{gd_{fr}}$.

The effective capacitance at node X1 is

$$\begin{aligned}C_{X1} = (C_{ds_{fr}} + C_{gd_{OV}} + C_{gd_{fr}})_{P2} + (C_{ds_{fr}} + C_{gd_{OV}} + C_{gd_{fr}})_{N1} + \\ (C_{gs})_{PINVL} + (C_{gs})_{NINVL}.\end{aligned} \quad (5.6)$$

Similarly, the capacitance at node X2 is

$$\begin{aligned}C_{X2} = (C_{ds_{fr}} + C_{gd_{OV}} + C_{gd_{fr}})_{P3} + (C_{ds_{fr}} + C_{gd_{OV}} + C_{gd_{fr}})_{N2} \\ + (C_{gs})_{PINVR} + (C_{gs})_{NINVR}.\end{aligned} \quad (5.7)$$

From Equations 5.4 and 5.6,

$$C_{X1(CCSA)} \gg C_{X1(CF-SA)}.$$

This instantaneously generates the potential at these nodes. Although P4 and P5 can be driven directly by nodes X1 and X2, the voltage difference at these two nodes is

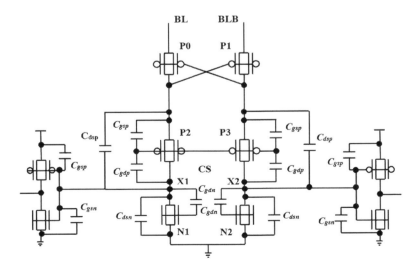

FIGURE 5.6 Capacitive model of current mode circuit in the current feed SA.

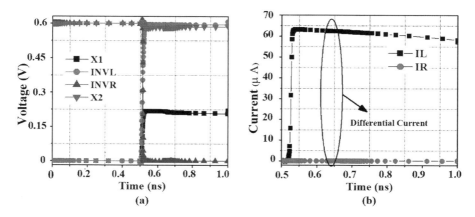

FIGURE 5.7 Signal waveforms: (a) Differential current in CF-SA. (b) Voltage waveforms at various nodes.

not sufficient to generate enough differential current. Hence, two inverters—INVL and INVR—are used instead. These inverters are configured to have low-V_{th} PFETs and high-V_{th} n-type field effect transistors (NFETs), generating complementary voltage levels. Finally, INVL and INVR drive the PFETs—P4 and P5—and feed large differential current as shown in Figure 5.7, thus developing large differential voltage on the datalines. After sufficient time, the voltage at DL/DLB becomes high, while I_{DIFF} is low, which reduces power dissipation by not allowing continuous current (I_{DIFF}) to flow. Figure 5.8 shows the voltage waveforms at various nodes from simulating the CF-SA scheme along with the voltage levels on the datalines (DL and DLB). Figure 5.8 is corresponding to read 0; in this case, the right branch passes

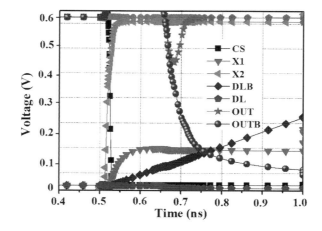

FIGURE 5.8 Simulated transient behavior of the CF-SA for V_{DD} = 0.6V and C_{BL} = 100fF.

more current (I_R) than the left (I_L) emphatically $V_{DLB} > V_{DL}$, as can be seen in Figure 5.8. The mismatch in CMC devices is crucial. Any mismatch in the CMC devices results in variation in the differential current. To ensure perfect matching of the CMC transistors, various techniques like the circuit's topology, proper sizing, and a symmetric layout design can be adopted. Since SRAM is an area-constraint design, adding additional devices for matching is not a preferable solution. Therefore, we appropriately upsized the CMC devices to minimize mismatch. The well-known Pelgrom model has demonstrated that the size-dependence mismatch is proportional to the inverse of the square root of the area, that is, $\sigma_{\Delta VTH} = \dfrac{A_{\Delta VTH}}{W.L}$. Here, the proportionally constant A_{AVT} characterizes the matching performance of the technology. Using this law, it is seen that the accuracy of a metal oxide semiconductor field effect transistor (MOSFET) can be increased by increasing its width or length.

5.6 FUNCTIONALITY AND DIFFERENTIAL VOLTAGE AND CURRENT ANALYSIS

During the standby mode, *BL* and DL are precharged to V_{DD} and ground, via the P_L–P_R and N3–N4 FETs, respectively. CS signal is at logic HIGH; this precharges nodes X1 and X2 to ground. Thus, the outputs of INVL and INVR become high, which keeps P4 and P5 in cutoff mode. Subsequently, the equalization device N_{EQ} is turned *ON* to clamp the nodes X3 and X4 to equal potential, while *PSEN* device is turned *OFF* to save power. During the read operation, wordline (WL) signal is triggered to HIGH to access a particular memory cell, creating differential current and voltage on bitline (*BL*). Concurrently, P2 and P3 are turned *ON* by asserting the CS signal to LOW. Owing to the large-sized *BL*-load FETs (P_L, P_R) and the P0, P1 of current mode logic, differential currents are quickly transported to the low-capacitive node X1, which, in turn, quickly generates the differential potential on these nodes. However, the voltage difference at these two nodes is not sufficient to generate enough differential voltage. Hence, two inverters—INVL and INVR— are used instead. These inverters are configured to have low-V_{th} PFETs and high-V_{th} NFETS, generating complementary voltage levels. The voltages at nodes X1 and X2 and the outputs of INVL and INVR are illustrated in Figure 5.7a. The INVL and INVR drive the PFETs—P4 and P5—and feed large differential current (refer to Figure 5.7b), thus developing a large differential voltage on the datalines. From a topological standpoint, this operation is more robust than that of a CCSA, because a large differential voltage is developed on data. There are two popular models for FinFET technology—predictive technology model (PTM) [26] and UFDG model [27]. We have used the double-gate PTM 22-nm technology model and specter as the platform for FinFET SA simulations. In our simulations, based on the equation $W_{eff} = (2.H_{FIN} + W_{FIN}) \times N_{FIN}$ (where fin height H_{FIN} and fin width W_{FIN} are fixed), we can change the number of fins—N_{FIN}—to get a reasonable effective channel width—W_{EFF}. Another advantage is that we need to select only minimum-sized transistors (P4, P5) to ensure reasonable differential current in CF-SA while maintaining leakage current and speed.

The five corners of the envelops that encompass all possible points are referred to as fast–fast (FF), fast–slow (FS), typical–typical (TT), slow–slow (SS), and slow–fast (SF). For example, an FF will refer to fast n-type metal-oxide semiconductor (NMOS) and fast PMOS transistors. Here, CF-SA_FF stands for the output current for current feed SA at the fast–fast process corner, which means fast PMOS and fast NMOS (low V_{th}). In particular, the corner simulation results for CCSA and CF-SA with varying V_{DD} are shown in Figure 5.9. It can be seen that CCSA is most critical at lower supply voltage. Furthermore, as the supply voltage decreases, the effect of improved differential current in CF-SA due to DL-decoupling scheme can be seen across all process corners. CF-SA topology develops more differential voltage, which directly translates to greater offset tolerance for GSA. Compared to CCSA, CF-SA provides on average 90% (92mV against 175mV) improvement in differential potential on DLs. A larger differential is able to tolerate larger GSA offset ΔV_{th} and offers lower cell access time. To put the comparative analysis in perspective, Figure 5.10 plots the simulated differential voltage distribution on the datalines as a result of equal cell current, for CCSA, has, and CF-SA. The CMC device sizes are kept identical for all simulations, and CF-SA. The CMC device sizes are kept identical for all simulations, and variations are applied to the primary (current) sensing stage only. A 1,500-point Monte Carlo (MC) analysis, considering V_{th} variation in transistors, is performed at the worst-case process corner. All circuits are simulated at a power supply of 0.6V, C_{DL} = 100fF, and a clock frequency of 1GHz. It is inferred from Figure 5.10 that CF-SA gives higher differential voltage, with 66.6% and 34.32% higher mean (μ) than CCSA and HSA, respectively. In our analysis, current ratio amplification (CRA) is defined as the ratio of actual differential current (output current of primary sensing stage) to I_{CELL} (i.e., $CRA = I_{DIFF}/I_{CELL}$). CRA is evaluated for CCSA, HSA, and CF-SA through successive simulations by varying the sizes of devices, and the results are illustrated in Figure 5.11. It is worth noting that CCSA is the most critical, since the cell current is close to the differential

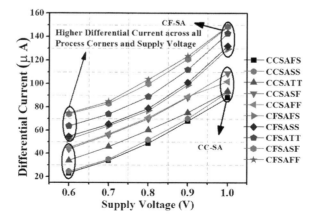

FIGURE 5.9 Simulated differential current versus supply voltage across all the process corners for CCSA and CF-SA.

Data Sensing in SRAM: A Hybrid Approach with FinFET

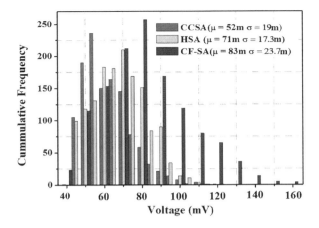

FIGURE 5.10 Differential voltage distribution of the designs in comparison for 1,500 MC simulations at room temperature.

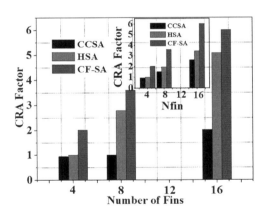

FIGURE 5.11 Current ratio amplification factor variation with a number of fins.

current, and thus, CRA is close to unity. For CF-SA, CRA is 2.12× and 1.99× higher than that in CCSA and HSA, respectively, for minimum device size and 0.6V V_{DD}. Figure 5.11 (inset) also verifies the higher CRA for CF-SA, at 1V V_{DD}.

5.7 DEVICE AND CAPACITANCE MISMATCH IN CF-SA

In Figure 5.12, the switching threshold voltage, V_M, is defined as the point where $V_{IN} = V_{OUT}$. Its value can be obtained graphically from the intersection of the voltage transfer characteristic (VTC) with the line given by $V_{IN} = V_{OUT}$ as shown in Figure 5.12. In this region, both PMOS and NMOS are always saturated, since

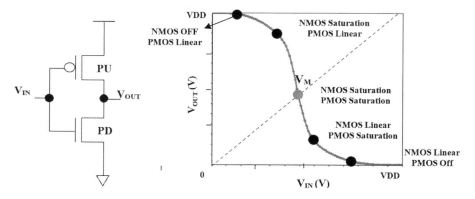

FIGURE 5.12 Inverter voltage transfer characteristic and switching threshold voltage.

$V_{DS} = V_{GS}$. An analytical expression for V_M is obtained by equating the currents through the transistors.

$$K_n V_{DSATn}\left(V_M - V_{thn} - \frac{V_{DSATn}}{2}\right) + K_p V_{DSATp}\left(V_M - V_{DD} - V_{thp}\frac{V_{DSATp}}{2}\right) = 0 \quad (5.8)$$

Solving for V_M yields

$$V_M = \frac{\left(V_{thn} + \frac{V_{DSATn}}{2}\right) + r\left(V_{DD} + V_{thp} + \frac{V_{DSATp}}{2}\right)}{1+r}. \quad (5.9)$$

Here, $r = \dfrac{k_p V_{DSATp}}{k_n V_{DSATn}}$, $k_p = \mu_p C_{ox} \left(W/L\right)_p$, and $k_n = \mu_n C_{ox} \left(W/L\right)_n$.

μ_p and μ_n is the hole and electron mobility, respectively. From Equation5. 9, we can derive the required ratio of PMOS versus NMOS transistor sizes such that the switching threshold is set to a desired value V_M. The effect of changing the W_p/W_n ratio is to shift the transient region of the VTC [28]. Increasing the width of the PMOS or the NMOS moves VM towards V_{DD} or *GND*, respectively. From Equation 5.9, it can be manifested that any mismatch in the NMOS and PMOS of the inverters results in different switching thresholds of the inverters (INV and INL in Figure 5.3). To minimize the effect of mismatch on switching threshold of the inverter, we upsized the inverter (INV and INL) devices. However, the device sizes have only a minor impact on the switching threshold of the inverter. Therefore, variations mostly cause a shift in the switching threshold, but the operation of the CF-SA is by no means affected.

Data Sensing in SRAM: A Hybrid Approach with FinFET

All the SA designs in the references list are based on the assumption that the datalines' capacitance is equal ($C_{DL} = C_{DLB}$). Therefore, the voltage difference developed on the DLs is governed by

$$V_{DL} = V_o + \frac{I_L \cdot \Delta t}{C_{DL}}$$

$$V_{DLB} = V_o + \frac{I_R \cdot \Delta t}{C_{DLB}}$$

$$\Delta V_{DL} = V_{DL} - V_{DLB} = \Delta t \cdot \frac{(I_L - I_R)}{C_{DL}} \text{ (For equal dataline capacitances)}$$

$$\Delta V_{DL} = \Delta t \cdot \frac{\Delta I}{C_{DL}}.$$

If CDL differs from CDLB, the SA works properly only if $I_L/C_{DL} > I_R/C_{DLB}$. Thus, it is important to note that in CCSA, HSA, and CF-SA, any mismatch in the C_{DL} and C_{DLB} affects the effective differential voltage on the datalines. Let us consider the read 1 case where I_L will be greater than I_R. In this case, the voltage at dataline DL (V_{DL}) will be greater than V_{DLB}. If some mismatch exists in the C_{DL} and C_{DLB} like $C_{DL} > C_{DLB}$, then V_{DL} (for $C_{DL} > C_{DLB}$) < V_{DL} (for $C_{DL} = C_{DLB}$) If the voltage difference for the two cases is $\Delta V_{DL}(Mismatch) = V_{DL1} - V_{DL2}$, here $V_{DL1} = V_{DL}$ (if $C_{DL} = C_{DLB}$). Similarly, $V_{DL2} = V_{DL}$ (if $C_{DL} > C_{DLB}$). Then $V_{DL}(Mismatch)$ will reduce, and thus, the effective differential voltage at dataline $\Delta V_{DL} = V_{DL} - V_{DLB}$ also will reduce. Thus, any mismatch in the C_{DL} and C_{DLB} kills the differential voltage on the datalines and reduces the read yield. As far as ΔV_{DL} is greater than the mismatch in V_{th} of the NMOS device (N_{DL} and N_{DR}), the GSA will latch in the correct direction. It is important to note that the GSA is shared among the number of columns. Therefore, the devices of the GSA are upsized to mitigate this concern.

To investigate the impact of a dataline capacitance mismatch, parametric variation simulations were conducted, and results are shown in Figures 5.13 and 5.14 for CCSA and CFSA, respectively. It can be manifested from Figure 5.13 that CCSA can tolerate up to 8fF mismatch and failed to resolve the data after this. However, CF-SA is more immune to C_{DL} mismatch results, and all outputs yield correct results as manifested in Figure 5.14. Figure 5.15 shows the output characteristics of CF-SA for C_{DL} mismatch to gauge the worst-case mismatch condition. CF-SA shows no failure up to 46fF of C_{DL} mismatch as illustrated in Figure. 5.15. Figure 5.16 shows the sensing delay characteristics of designs for C_{DL} mismatch; it can be noted that CCSA can only tolerate up to an 8% difference between the two DL capacitances with higher sensing delay. The CF-SA, in contrast, is less sensitive to dataline capacitance mismatch due to its dataline decoupled scheme with minimum sensing delay variations. Thus, CF-SA is more robust against C_{DL} mismatch.

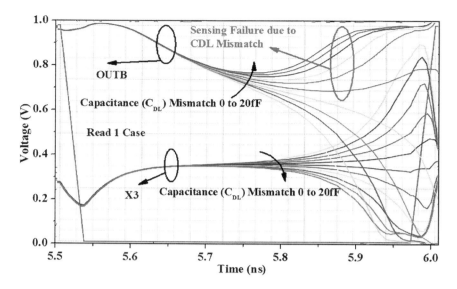

FIGURE 5.13 Output characteristics of CCSA versus mismatch between C_{DL} and C_{DLB}.

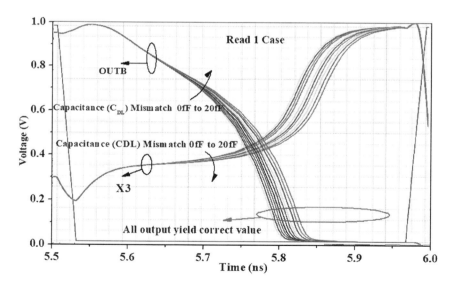

FIGURE 5.14 Output characteristics of CF-SA versus mismatch between C_{DL} and C_{DLB}.

Data Sensing in SRAM: A Hybrid Approach with FinFET

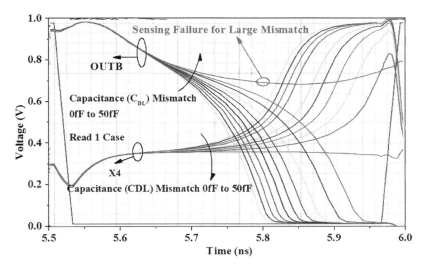

FIGURE 5.15 Output characteristics of CF-SA versus mismatch between C_{DL} and C_{DLB}.

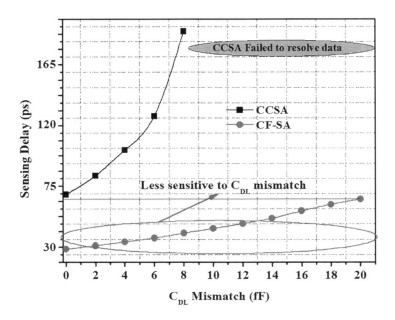

FIGURE 5.16 Sensing delay variation with C_{DL} mismatch.

5.8 OFFSET ANALYSIS

For high resolution, it is important to minimize the SA's input-referred offset voltage V_{OS}, which is largely determined by a threshold mismatch of the sensing transistors. The simplest way to reduce the SA's V_{OS} is to increase the size of the devices [24]; one approach is to add devices to provide a feedback mechanism to reduce the SA's sensitivity to V_{th} mismatches. Alternatively, additional circuitry can be used to calibrate the offset, either dynamically or with post-process trimming [29]. Another approach is using multistage timing in which the connections between the SAs are changed to reduce V_{th} mismatches. However, the consequences of all these approaches are increased die area, BL loading, and power dissipation. In CCSA, HSA, and CF-SA, V_{DIFF} is the voltage created by I_{CELL}, which must be larger than the sum of counteracting offset due to GSA. Since the proposed scheme produces a large differential at the input of GSA, it is more immune to offset issues.

Initially, simulations were carried out to determine the effect of offset in the input transistors by observing the output nodes—OUT and OUTB. To introduce an offset during initial simulations, a V_{th} adder was used and variations from 10 to 110mV were made. A standard 6T cell with the same I_{CELL} was used for all simulations. We define V_{th} adder as the amount by which the offset is shifted. Simulations were conducted under several offset values in latch GSA for all schemes. Figure 5.17a shows that a correct decision was made for V_{th} mismatch up to 40mV for CCSA, while HSA was able to sense the correct data up to 50mV of V_{th} mismatch in input devices

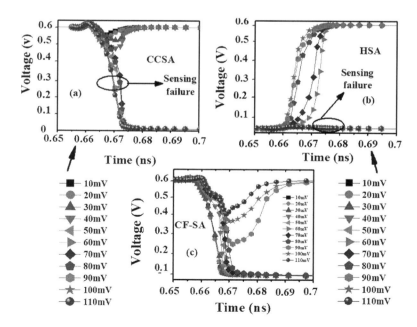

FIGURE 5.17 Offset simulation results for designs in comparison: (a) CCSA, (b) HAS, and (c) CF-SA.

Data Sensing in SRAM: A Hybrid Approach with FinFET 169

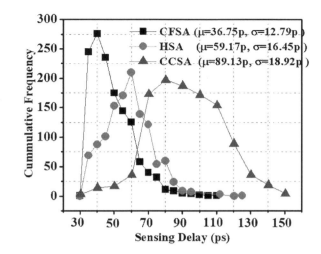

FIGURE 5.18 Sensing delay distributions for the designs in comparison, results are from a 1,500-sample MC simulation that varies local parameters (6σ) at room temperature and $C_{BL} = C_{DL} = 100\text{fF}$.

(Figure 5.17b). On the other hand, CF-SA made a correct decision up to an 80mV V_{th} mismatch (Figure 5.17c). Since more differential appears on DLs in CF-SA, it is more immune to the inherent offset in GSA. In this work, we investigate the input-offset quality of the SA designs. Therefore, variation investigations are made for SAs alone. Circuit characteristics with intra-die variation, capturing the local mismatch behavior and process variation, obtained by MC simulation of 1,500 iterations, are shown in Figure 5.18. The transistor variability used in the MC simulation was modeled as a V_{th} variation that follows a Gaussian distribution. As observed in Figure 5.18, sensing delay distribution curves for the CCSA and CF-SA cross at 67ps. Based on the simulation data, our estimation shows that 97.4% of statistical samples in the case of CF-SA have a sensing delay lower than 67 ps, signifying its lower sensing delay compared to CCSA (69.6% of statistical samples in the case of CCSA have sensing delay higher than 67 ps). Similarly, 62.06% of statistical samples in the case of CF-SA have sensing delay lower than 52 ps, signifying its lower sensing delay compared to HSA (27.4% of statistical samples in the case of HSA have a sensing delay lower than 52 ps). The CF-SA architecture shows less sensing delay and variability and is well justified with the measured mean (μ) and standard deviation (σ) as presented in Table 5.1. Significantly, for CF-SA, 32.39% and 22.24% reductions in σ_{Delay} than CCSA and HSA are achieved, respectively, for a 0.6V supply voltage with optimally sized devices.

Figure 5.19 shows how a CF-SA layout can be built using FinFETs [30–31]. The approximated circuit layout of CFSA was carried out according to 20-nm logic-based design rules, since PTM does not give any information about layout design. The CFSA, CCSA, and HSA follow the same bitline pitch. The resulting layout,

TABLE 5.1
Summary of All Designs in Comparison with CF-SA in Terms of Mean (μ) and Standard Deviation (σ) for $C_{DL} = 100fF$

V_{DD} (V)		CCSA	HSA	CF-SA
0.6	μ (mV)	89.13	59.17	38.75
	σ (mV)	18.92	16.45	12.79
0.7	μ (mV)	66.27	53.12	33.7
	σ (mV)	16.10	11.95	8.81
0.8	μ (mV)	51.96	41.18	26.10
	σ (mV)	11.97	8.92	7.89
0.9	μ (mV)	42.16	34.1	20.81
	σ (mV)	9.37	6.89	5.71
1.0	μ (mV)	39.1	27.9	18.1
	σ (mV)	7.46	4.86	3.32

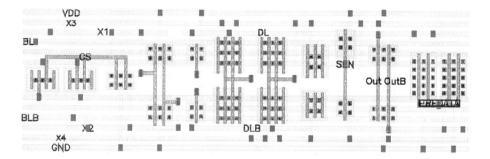

FIGURE 5.19 Physical design view of current feed sense amplifier.

shown in Figure 5.22, fits into the 8.94μm² footprint. The layout is carefully designed so that sensitive pairs do not come in close proximity. Symmetric layout is a critical piece of our design strategy to reduce the effect of mismatch.

5.9 PERFORMANCE AND POWER MEASUREMENT

This section of the thesis discusses the efficiency of CF-SA and its design tradeoffs and makes a quantitative comparison between CF-SA, CCSA, and HSA. The worst-case corner simulations are performed for all designs to verify the results and compare the performance of CF-SA with that of CCSA and HAS. Simulations are conducted at typical (TT, 27°C), worst-power (FF, 240°C) and worst-speed (SS, 85°C) conditions. In this comparative analysis, I_{CELL} = 32.02μA, V_{DD} = 0.6V, C_{DL} = 100fF, and there are 128 cells per column. Comparing the conventional schemes, all

Data Sensing in SRAM: A Hybrid Approach with FinFET 171

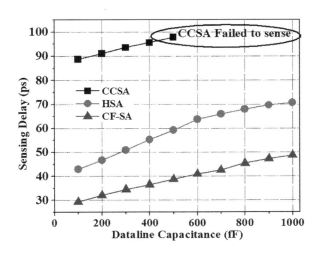

FIGURE 5.20 Sensing delay variation with dataline capacitance.

three schemes employ an identical latch as GSA. Read performance is evaluated in terms of sensing delay that is the amount of time taken by the node that fully discharges to reach 50% of V_{DD} from the moment the CS signal rises to 50%. Dataline capacitance has a greater impact on the performance of SAs with a higher slope along the dataline capacitance axis. Hence, the DL's capacitance is the bottleneck for speed and correct sensing. In order to gauge the effect of huge DLs' capacitance, extensive simulations were carried out with varying DLs' capacitance. Figure 5.20 demonstrates the superiority of CF-SA over other designs at 0.6V supply voltage, against C_{DL} variation. This shows that CF-SA offers enhanced speed robustness against varying C_{DL}, giving a delay sensitivity of only 1.9ps/100fF, which is better than that of HSA at 2.78ps/100fF. Although CCSA follows the same trend, it failed to sense after 500fF because of lower I_{CELL}.

Figure 5.21 illustrates power dissipation during read and static power dissipation at different supply voltages. The energy consumption for CF-SA is less than that of CCSA due to low power consumption during the amplification stage (GSA). Energy consumption in CF-SA is also independent of the bitline capacitance as maintains the bitline's voltage close to V_{DD}. Results indicate that CF-SA offers 35.18% saving in energy consumption over CCSA for 128 cells per column and C_{DL} = 100fF. By comparison, the HSA design consumes 56.59% less power than CF-SA owing to its better precharing scheme. Figure 5.21 (inset) shows the static power consumption for all designs being compared. All three SAs are turned off by setting their control signals to either ground or V_{DD}, and V_{DD} is swept from 0.5 to 1V so as to cover the commercial standard range. While Figure 5.21 suggests that CF-SA consumes high standby power (due to more leakage components), it is still comparable to its CCSA counterpart. The two internal inverters (INVL and INVR) have more impact on leakage current in CF-SA.

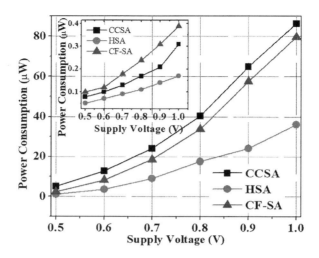

FIGURE 5.21 Performance comparison of CF-SA with respect to CCSA and HSA, in terms of read and hold power (inset).

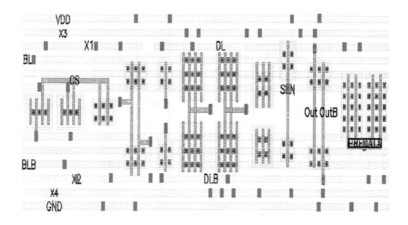

FIGURE 5.22 Layout view of current feed SA.

5.10 CHAPTER SUMMARY

In this chapter, for the first time, a FinFET current mode sense amplifier, targeting small I_{CELL} sensing in SRAM, was presented with a detailed sensitivity analysis. The sense amplifier was implemented with a 22-nm FinFET technology for high read yield and low power applications. The proposed CF-SA employed a dataline-isolated current sensing approach, thus enabling CF-SA to achieve 3.02× and 1.45× faster read speeds than CCSA and HSA, respectively, when sensing 32.02μA on a 100fF dataline. It was shown that CF-SA generates 2.23× and 1.7× more differential voltage on datalines than CCSA and HSA, respectively. Based on simulation results, we can conclude that, compared to other current mode FinFET SAs, during the read operation, CF-SA with isolated datalines approach has such advantages as robust read access against small I_{CELL} a huge dataline capacitance and low sensing delay. Furthermore, CF-SA offers greater offset tolerance (up to 80mV). The effect of the number of fins on differential current and voltage was also compared.

REFERENCES

1. S. Ataei and J. E. Stine, "A 64 kB Approximate SRAM Architecture for Low-Power Video Applications," *IEEE Embedded Systems Letters*, vol. 10, no. 1, pp. 10–13, March 2018, doi:10.1109/LES.2017.2750140.
2. D. Patel, A. Neale, D. Wright and M. Sachdev, "Body Biased Sense Amplifier with Auto-Offset Mitigation for Low-Voltage SRAMs," *IEEE Transactions on Circuits and Systems I: Regular Papers*, vol. 68, no. 8, pp. 3265–3278, August 2021, doi:10.1109/TCSI.2021.3081917.
3. C. Hsieh, M. Fan, V. P. Hu, P. Su and C. Chuang, "Independently-Controlled-Gate FinFET Schmitt Trigger Sub-Threshold SRAMs," *IEEE Transactions on Very Large Scale Integration (VLSI) Systems*, vol. 20, no. 7, pp. 1201–1210, July 2012, doi:10.1109/TVLSI.2011.2156435.
4. N. Maroof and B. Kong, "10T SRAM Using Half-VDD Precharge and Row-Wise Dynamically Powered Read Port for Low Switching Power and Ultralow RBL Leakage," *IEEE Transactions on Very Large Scale Integration (VLSI) Systems*, vol. 25, no. 4, pp. 1193–1203, April 2017, doi:10.1109/TVLSI.2016.2637918.
5. Y. Lai and S. Huang, "A Resilient and Power-Efficient Automatic-Power-Down Sense Amplifier for SRAM Design," *IEEE Transactions on Circuits and Systems II: Express Briefs*, vol. 55, no. 10, pp. 1031–1035, October 2008, doi:10.1109/TCSII.2008.926797.
6. C.-X. Xue, W.-C. Zhao, T.-H. Yang, Y.-J. Chen, H. Yamauchi and M.-F. Chang, "A 28-nm 320-kb TCAM Macro Using Split-Controlled Single-Load 14T Cell and Triple-Margin Voltage Sense Amplifier," *IEEE Journal of Solid-State Circuits*, vol. 54, no. 10, pp. 2743–2753, October 2019, doi:10.1109/JSSC.2019.2915577.
7. M.-F. Chang et al., "A Sub-0.3 V Area-Efficient L-Shaped 7T SRAM with Read Bitline Swing Expansion Schemes Based on Boosted Read-Bitline, Asymmetric-V_{th} Read-Port, and Offset Cell VDD Biasing Techniques," *IEEE Journal of Solid-State Circuits*, vol. 48, no. 10, pp. 2558–2569, October 2013, doi:10.1109/JSSC.2013.2273835.
8. Y. Chen et al., "Compact Measurement Schemes for Bit-Line Swing, Sense Amplifier Offset Voltage, and Word-Line Pulse Width to Characterize Sensing Tolerance Margin in a 40 nm Fully Functional Embedded SRAM," *IEEE Journal of Solid-State Circuits*, vol. 47, no. 4, pp. 969–980, April 2012, doi:10.1109/JSSC.2012.2185180. [From Thesis]
9. W. Zhu et al., "A Wide-Range-Supply-Voltage Sense Amplifier Circuit for Embedded Flash Memory," *IEEE Transactions on Circuits and Systems II: Express Briefs*, vol. 67, no. 8, pp. 1454–1458, August 2020, doi:10.1109/TCSII.2019.2939160.
10. G. D. Licciardo, L. Di Benedetto, A. De Vita, A. Rubino and A. Femia, "A Bit-Line Voltage Sensing Circuit with Fused Offset Compensation and Cancellation Scheme," *IEEE Transactions on Circuits and Systems II: Express Briefs*, vol. 66, no. 10, pp. 1633–1637, October 2019, doi:10.1109/TCSII.2019.2928456.
11. S. Cosemans, W. Dehaene and F. Catthoor, "A 3.6 pJ/Access 480 MHz, 128 kb On-Chip SRAM with 850 MHz Boost Mode in 90 nm CMOS with Tunable Sense Amplifiers," *IEEE Journal of Solid-State Circuits*, vol. 44, no. 7, pp. 2065–2077, July 2009, doi:10.1109/JSSC.2009.2021925.

12. M. E. Sinangil and A. P. Chandrakasan, "Application-Specific SRAM Design Using Output Prediction to Reduce Bit-Line Switching Activity and Statistically Gated Sense Amplifiers for Up to 1.9$\times$ Lower Energy/Access," *IEEE Journal of Solid-State Circuits*, vol. 49, no. 1, pp. 107–117, January 2014, doi:10.1109/JSSC.2013.2280310.
13. M. Sharifkhani, E. Rahiminejad, S. M. Jahinuzzaman and M. Sachdev, "A Compact Hybrid Current/Voltage Sense Amplifier with Offset Cancellation for High-Speed SRAMs," *IEEE Transactions on Very Large Scale Integration (VLSI) Systems*, vol. 19, no. 5, pp. 883–894, May 2011, doi:10.1109/TVLSI.2009.2039949.
14. D. Anh-Tuan, K. Zhi-Hui and Y. Kiat-Seng, "Hybrid-Mode SRAM Sense Amplifiers: New Approach on Transistor Sizing," *IEEE Transactions on Circuits and Systems II: Express Briefs*, vol. 55, no. 10, pp. 986–990, October 2008, doi:10.1109/TCSII.2008.2001965.
15. M. Chang and S. Shen, "A Process Variation Tolerant Embedded Split-Gate Flash Memory Using Pre-Stable Current Sensing Scheme," *IEEE Journal of Solid-State Circuits*, vol. 44, no. 3, pp. 987–994, March 2009, doi:10.1109/JSSC.2009.2013763.
16. C. D. Matthus, S. Buhr, M. Kreißig and F. Ellinger, "High Gain and High Bandwidth Fully Differential Difference Amplifier as Current Sense Amplifier," *IEEE Transactions on Instrumentation and Measurement*, vol. 70, pp. 1–11, 2021, Art no. 2000911, doi:10.1109/TIM.2020.3018830.
17. S. Choi et al., "A Self-Biased Current-Mode Amplifier with an Application to 10-bit Pipeline ADC," *IEEE Transactions on Circuits and Systems I: Regular Papers*, vol. 64, no. 7, pp. 1706–1717, July 2017, doi:10.1109/TCSI.2017.2676105.
18. C. J. Chevallier et al., "A 0.13μm 64Mb Multi-Layered Conductive Metal-Oxide Memory," *2010 IEEE International Solid-State Circuits Conference (ISSCC)*, pp. 260–261, 2010, doi:10.1109/ISSCC.2010.5433945.
19. R. Pușcașu, P. Brînzoi, L. Creoșteanu and G. Brezeanu, "A High Voltage Current Sense Amplifier with Extended Input Common Mode Range Based on a Low Voltage Operational Amplifier Cell," *2015 European Conference on Circuit Theory and Design (ECCTD)*, pp. 1–4, 2015, doi:10.1109/ECCTD.2015.7300078.
20. E. Seevinck, P. J. van Beers and H. Ontrop, "Current-Mode Techniques for High-Speed VLSI Circuits with Application to Current Sense Amplifier for CMOS SRAM's," *IEEE Journal of Solid-State Circuits*, vol. 26, no. 4, pp. 525–536, April 1991, doi:10.1109/4.75050.
21. M. Fan, V. P. Hu, Y. Chen, P. Su and C. Chuang, "Variability Analysis of Sense Amplifier for FinFET Subthreshold SRAM Applications," *IEEE Transactions on Circuits and Systems II: Express Briefs*, vol. 59, no. 12, pp. 878–882, December 2012, doi:10.1109/TCSII.2012.2231016.
22. S. Mukhopadhyay, H. Mahmoodi and K. Roy, "A Novel High-Performance and Robust Sense Amplifier Using Independent Gate Control in Sub-50-nm Double-Gate MOSFET," *IEEE Transactions on Very Large Scale Integration (VLSI) Systems*, vol. 14, no. 2, pp. 183–192, February 2006, doi:10.1109/TVLSI.2005.863743.
23. J. P. Kulkarni et al., "5.6 Mb/mm² 1R1W 8T SRAM Arrays Operating Down to 560 mV Utilizing Small-Signal Sensing with Charge Shared Bitline and Asymmetric Sense Amplifier in 14 nm FinFET CMOS Technology," *IEEE Journal of Solid-State Circuits*, vol. 52, no. 1, pp. 229–239, January 2017, doi:10.1109/JSSC.2016.2607219.
24. A. Fritsch et al., "24.1 A 6.2 GHz Single Ended Current Sense Amplifier (CSA) Based Compileable 8T SRAM in 7nm FinFET Technology," *2021 IEEE International Solid-State Circuits Conference (ISSCC)*, pp. 334–336, 2021, doi:10.1109/ISSCC42613.2021.9365812.
25. M. Tsai, J. Tsai, M. Fan, P. Su and C. Chuang, "Variation Tolerant CLSAs for Nanoscale Bulk-CMOS and FinFET SRAM," *2012 IEEE Asia Pacific Conference on Circuits and Systems*, pp. 471–474, 2012, doi:10.1109/APCCAS.2012.6419074.
26. PTM-MG Multi-Gate Model for Multi-Gate FinFET Transistors (2013). http://ptm.asu.edu/.

27. SOI Group of University of Florida, UFDGMOSFET Model User Guide (Linux Version) (2003). www.soi.tec.ufl.edu.
28. N. H. E. Waste and D. M. Harris, "CMOS VLSI Design," 2011. http://fa.ee.sut.ac.ir/Downloads/AcademicStaff/24/Courses/11/CMOS%20VLSI%20Design%20A%20Circuits%20and%20Systems%20Perspective,%204th%20Edition%20(2011).pdf.
29. M. Bhargava, M. P. McCartney, A. Hoefler and K. Mai, "Low-Overhead, Digital Offset Compensated, SRAM Sense Amplifiers," *2009 IEEE Custom Integrated Circuits Conference*, pp. 705–708, 2009, doi:10.1109/CICC.2009.5280732.
30. M. Alioto, "Analysis of Layout Density in FinFET Standard Cells and Impact of Fin Technology," *Proceeding of IEEE International Symposium on Circuits and Systems (ISCAS)*, Paris, France, 2010, pp. 3204–3207, doi:10.1109/ISCAS.2010.5537930.
31. J. Ryckaert et al., "Design Technology Co-Optimization for N10," *Proceeding of IEEE, Custom Integrated Circuit Conference (CICC)*, San Jose, CA, 2014, pp. 1–8, doi:10.1109/CICC.2014.6946037.

6 BTI-Aware and Soft-Error-Tolerant SRAM

6.1 INTRODUCTION

Due to the rapid scaling of integrated devices and circuits, reliability issues have become more serious concerns in modern applications [9]. Furthermore, with the scaling of supply voltages, the node capacitance reduces, which makes the circuit susceptible to external noise caused by alpha particles and cosmic rays. The external noise sources can trigger a glitch at the sensitive nodes of a certain circuit, which gives rise to so-called **transient faults**. If such a fault occurs at the storage node of memory elements, it can alter the respective logic state, and a **soft error** can occur [7]. Soft errors due to radiations such as cosmic rays and alpha particles are a significant concern in modern complementary metal-oxide semiconductor (CMOS) technologies [22]. SRAM cells are particularly susceptible to soft errors because the critical charge (Q_{crit}) required to upset a cell is exceptionally small due to the typically low supply voltages, small noise margins, and lower node capacitances [11]. Because of this, the static random-access memory (SRAM) cells are affected by the external particle radiations that cause the soft error [13,15]. The **single-event upset (SEU)** induced by radiation particles in terrestrial and aerospace applications is the major failure mechanism that causes the failure of the electronics systems by temporarily flipping the stored data [6,12]. When the sensitive node of a circuit is hit by the high-energy particle, the induced charge is collected and accumulated through the drift process. Once the generated voltage pulse from the accumulated charge is above the switching threshold, the stored data of the sensitive node flips [10,27].

Next to soft errors, the performance of single metal-oxide semiconductor (MOS) devices used in SRAM cells is additionally affected by aging mechanisms, which typically have a detrimental impact on the stability of SRAM cells. The primary aging mechanisms for modern integrated devices are the so-called bias temperature instabilities (BTIs). BTIs can be classified into positive BTIs (PBTIs) and negative BTIs (NBTIs), depending on whether a positive or a negative bias is applied at the gate of the transistors. Both PBTIs and NBTIs can be observed for n-type and p-type metal-oxide semiconductor (NMOS and PMOS, respectively) transistors. In general, BTI affects the performance of MOS transistors when they are biased in strong inversion [20]. For instance, the absolute value of the threshold voltage of PMOS and NMOS transistors drifts toward larger values with the stress time, as shown in Figure 6.1 for 32-nm CMOS technology using HSPICE. As can be seen, the effect of NBTIs in PMOS transistors is more severe than PBTIs in NMOS transistors. The observed increase in the threshold voltage of transistors can lead to a decrease in the stability of SRAM cells and can increase the overall delay of the circuit. Even a small variation of the threshold voltage can affect the noise margin of the SRAM cells.

BTI-Aware and Soft-Error-Tolerant SRAM 177

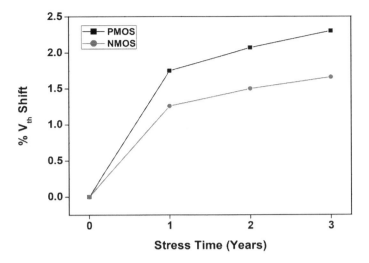

FIGURE 6.1 Threshold voltage degradation (percentage of initial threshold voltage) for the PMOS/NBTI and NMOS/PBTI case as predicted by the HSPICE for a 32-nm technology node considering DC stress. The rate of threshold voltage degradation is high for the initial years and tends to saturate over time.

6.2 RADIATION HARDENING ANALYSIS METHODOLOGY

The radiation hardening investigation of the various previously mentioned SRAM cells is carried out using the HSPICE tool. Figure 6.2 summarizes the various steps involved in the critical charge analysis with various supply voltages and operating temperatures for different SRAM cells.

6.2.1 CRITICAL CHARGE

To inject SEU into the simulated circuits, the current induced by α-particles hitting CMOS circuits is modeled by a double exponential current source specified by Equation 6.1 [14].

$$I_{inj}(t) = \frac{Q_{inj}}{\tau_f - \tau_r} \times (e^{-t/\tau_f} - e^{-t/\tau_r}) \quad (6.1)$$

$$I_{inj}(t) = I_{peak} \times (e^{-t/\tau_f} - e^{-t/\tau_r}) \quad (6.2)$$

Q_{inj} is the total amount of charge deposited at the sensitive node, and I_{peak} is the peak value of the equivalent current source. The characteristic time constants τ_f and τ_r are material-dependent parameters. As suggested by Alouani et al. [3], we have used typical values of $\tau_f = 1$ ps and $\tau_r = 50$ ps. In our simulations, the critical

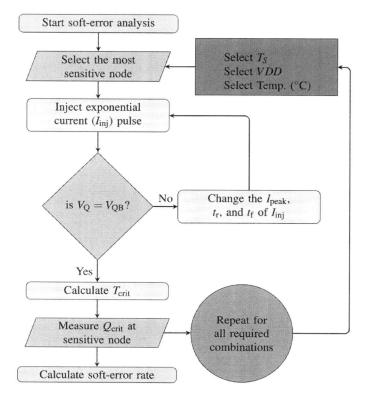

FIGURE 6.2 Simulation flow of soft-error analysis with different operating temperatures and supply voltages with embedded aging effects.

charge Q_{crit} is determined by injecting the current peak at the sensitive node of the respective SRAM cell, as shown in Figure 6.3, which simulates a possible particle strike.

In our experiments, the Q_{crit} is determined by injecting current at the sensitive node of the SRAM cells as shown in Figure 6.3. This pulse simulates the current induced by the particle strike at the sensitive node. To calculate Q_{crit}, we determine the minimum magnitude and duration of injected current pulse that is sufficient to flip the state of the storage node. Hence, Q_{crit} is determined by integrating the current pulse for the time interval 0 to T_{crit}. Therefore, the injected charge until T_{crit} is sufficient to make $V_Q = V_{QB}$. Hence, the critical charge can be given as

$$Q_{crit} = \int_0^{T_{crit}} I_{inj}(t)dt, \qquad (6.3)$$

where $I_{inj}(t)$ is the injected current pulse at the sensitive node for the SEU analysis.

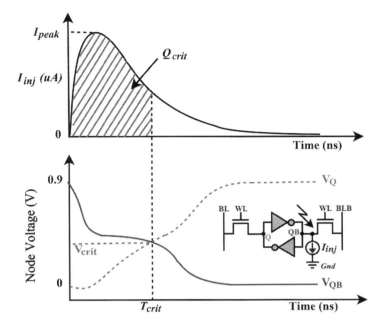

FIGURE 6.3 Graphical representation defining the critical charge. (top) Shows the transient behavior of the current pulse representing a possible particle strike. (bottom) V_Q and V_{QB} are the storage node voltages of the SRAM cell. The simulation setup for a high-energy particle hitting a sensitive node V_{QB} storing '1' of an SRAM cell is given in the inset.

6.2.2 Soft-Error Rate

A soft error in any digital circuit occurs when a high-energy particle from chip packaging materials or cosmic radiation induces enough charge that, if it is collected by a sensitive node, can change its logic value of the hit circuit node. Here, the soft-error rate (SER) of circuits is analyzed by considering memory cells. Circuits designed using deep submicron technology are highly susceptible to soft errors due to small critical charge attributed to the lower supply voltages and smaller node capacitances. The soft error resilience of any circuit can be evaluated from its Q_{crit}. The SER has an exponential dependency on Q_{crit}, and it is observed that the higher value of Q_{crit} translates into a lower SER [5]. The SER can be expressed as given by Equation 6.4:

$$SER \propto N_{\text{flux}} A e^{-\frac{Q_{\text{crit}}}{Q_S}}, \qquad (6.4)$$

where N_{flux} is the neutron flux intensity, A is the cross-sectional area of the sensitive node, and Q_S is the charge collection efficiency of the device in fC. From Equation 6.4, it is observed that a small increase in Q_{crit} will significantly reduce the SER.

6.2.2.1 Intra-Cell SERR

For the validation of soft-error hardening enhancement on different SRAM cells with the supply voltage and temperature variations, we have computed the intra-cell **SER ratio (SERR)**. The intra-cell SERR provides information about the soft error hardening of a particular SRAM cell with supply voltage and temperature variations. The approximate SERR for subthreshold and super-threshold operations ($SERR_V$) is introduced to analyze the effect of supply variations, and it can be calculated by assuming all other parameters unaffected except Q_{crit} for a particular operating temperature. Hence,

$$SERR_V = \left.\frac{SER_{superthreshold}}{SER_{subthreshold}}\right|_{@T} = \left.\frac{SER_{0.9V}}{SER_{0.5V}}\right|_{@T}. \quad (6.5)$$

Or

$$SERR_V \approx Antilog_e \left[Q_{crit}^{@0.5V} - Q_{crit}^{@0.9V}\right]_{@T}, \quad (6.6)$$

where $SER_{0.5V}$ and $SER_{0.9V}$ are the SERs and $Q_{crit}^{@0.5V}$ and $Q_{crit}^{@0.9V}$ are the critical charges at supply voltages of 0.5V and 0.9V, respectively. Similarly, the approximate SERR for room and elevated temperature ($SERR_T$) is also introduced to analyze the effect of temperature variations and it can be calculated by assuming all other parameters unaffected except Q_{crit} for particular supply voltage. Hence,

$$SERR_T = \left.\frac{SER_{25°C}}{SER_{125°C}}\right|_{@V_{DD}}. \quad (6.7)$$

Or

$$SERR_T \approx Antilog_e \left[Q_{crit}^{@125°C} - Q_{crit}^{@25°C}\right]_{@V_{DD}}, \quad (6.8)$$

where $SER_{25°C}$ and $SER_{125°C}$ are the SERs and $Q_{crit}^{@25°C}$ and $Q_{crit}^{@125°C}$ are the critical charges at operating temperatures of 25°C and 125°C, respectively. A lower SERR means better radiation hardening enhancement of the design considering supply voltage and temperature variations.

6.2.2.2 Inter-Cell SERR

An inter-cell SERR is calculated to analyze the soft error rate of the SRAM cell as compared to a 6T SRAM cell for the extreme conditions of the supply voltages and operating temperatures. The approximate inter-cell SERR ($SERR_\#$) for the considered SRAM cell normalized to 6T SRAM cell is introduced, and it can be calculated by assuming all other parameters unaffected except Q_{crit} for particular operating temperature and supply voltage. Hence,

$$SERR_{\#} = \left.\frac{SER_{\#}}{SER_{6T}}\right|_{@\text{VDD and T}}.\quad(6.9)$$

Or

$$SERR_{\#} \approx \left.\text{Antilog}_e\left[Q_{\text{crit}}^{6T} - Q_{\text{crit}}^{\#}\right]\right|_{@\text{VDD and T}},\quad(6.10)$$

where $SER_{\#}$ and SER_{6T} are the SERs and $Q_{\text{crit}}^{\#}$ and Q_{crit}^{6T} are the critical charges for the considered SRAM cell and reference 6T SRAM cell, respectively. Here, # indicates the considered SRAM cell for which SERR needs to be analyzed. The lower the $SERR_{\#}$ of any SRAM cell means that the soft error has less of an effect on the considered SRAM cell.

6.3 STABLE AND RELIABLE SRAM CELLS

In recent investigations where the conventional 6T SRAM cell, as shown in Figure 6.4a, has been analyzed, it has been observed that this cell does not provide adequate immunity against reliability issues, especially in harsh environments [11]. The 6T SRAM cell exhibits a reduced critical charge, which is not suitable for applications in which perturbations due to high-energy particles are more likely.

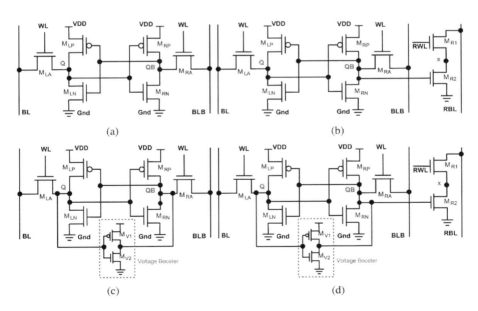

FIGURE 6.4 Schematic of different SRAM Cells: (a) 6T SRAM cell, (b) read-decoupled 8T (RD8T) SRAM cell after (c) asymmetric radiation-hardened 8T (AS8T) SRAM cell after (d) asymmetric radiation-hardened (AS10T) SRAM cell with improved stability.

To circumvent these limitations, Chang et al. [4] proposed read-decoupled 8T SRAM cell (RD8T), which requires separate write and read wordlines and bitlines, as shown in Figure 6.4b. This cell improves the read static noise margin (RSNM) to be approximately equal to the hold static noise margin (HSNM), without affecting other cell properties. The storage node capacitance of the RD8T SRAM cell remained the same as for the 6T SRAM cell. Although the RSNM was improved, the RD8T cell is still affected by external radiation at the same level of sensitivity as the 6T SRAM cell; thus, no improvement in the soft-error hardening could be achieved. Alouani et al. [3] discussed an asymmetric SRAM cell (AS8T) to improve soft-error hardening. As shown in Figure 6.4c, a minimum-sized charge booster is connected between the storage nodes to increase the node capacitance and thus improve the soft-error hardening. However, the RSNM of this SRAM cell remained at the same level as for the 6T SRAM cell. A pseudo-differential SEU-immune 12T (PD12T) SRAM cell has been presented by Ahmad et al. [2]. This cell exhibits soft-error hardening but requires a large area. The we-Quatro circuit is one of the promising approaches providing robust operation under a strongly radiative environment. However, the large area overhead might lead to a higher write failure rate due to process variations [15]. A PPN-based 10T SRAM cell with better performance has been discussed by Sanvale et al. [21], but its soft-error tolerance has not been investigated so far. Based on the preceding discussion, there is a high demand for optimizing SRAM cells, improving their soft-error resilience, and improving the RSNM while avoiding an increase in the required chip area and ensuring high signal integrity and low power consumption of the SRAM cell.

6.3.1 RADIATION-HARDENED ASYMMETRIC 10T SRAM CELL

In this section, an asymmetric radiation-hardened 10T (AS10T) SRAM cell is discussed, and the corresponding circuit is shown in Figure 6.4d. The AS10T SRAM cell is designed to improve the soft-error performance at the minimum possible area overhead. To create the AS10T SRAM cell, a minimum-sized CMOS inverter (M_{V1} and M_{V2}, named a charge booster) is placed between the storage nodes of the RD8T SRAM cell. The proposed design is inspired by the RD8T SRAM cell and is targeted to achieve an increased RSNM because of the separate read-decoupled circuit (M_{R1} and M_{R2}) during the read operation. Furthermore, the design is inspired by the AS8T SRAM cell to enhance the soft-error hardening at the storage node by connecting a charge booster circuit. The newly proposed AS10T cell combines the advantages of the RD8T and the AS8T SRAM cells. In the AS10T cell, the role of the charge booster is to withstand the impact of particle strikes at the storage nodes of the SRAM cell. The charge booster increases the storage node capacitance and recovers the data by pulling back the correct logic state. The control signals for different operations of the proposed AS10T SRAM cell are given in Table 6.1. In the following, the soft-error-hardening enhancement is thoroughly analyzed with consideration of the aging of MOS devices employing different SRAM cells as well as the proposed AS10T SRAM cell. The results reveal that the AS10T SRAM cell exhibits an overall enhanced soft error immunity because of an increased minimum amount of charge required to flip the

BTI-Aware and Soft-Error-Tolerant SRAM

TABLE 6.1
Control Signals of Proposed AS10T SRAM Cell for Different Operations

Operations	Control signals				
	BL	BLB	WL	RWL	RBL
Write 0	0	1	1	0	0
Write 1	1	0	1	0	0
Hold	×	×	0	0	0
Read	1	×	1	1	1

Write '0' and write '1' correspond to the storage node Q. *BL* and *BLB* are the bitlines and are complementary to each other. RBL is the bitline during the read operation. WL and RWL are the wordlines during the write and read operations, respectively.

stored data. The level of reliability in the proposed cell depends on the direction of the charge booster connection.

6.4 LEAKAGE CURRENT ESTIMATION

The leakage current during the HOLD operation of the SRAM cell is the major issue in the scaled CMOS technology. The leakage current is the major contributor to the total power dissipation of the circuit, especially during the HOLD operation. In any of the SRAM cells, subthreshold leakage current (I_{sub}), junction leakage current (I_{jn}), and gate leakage current (I_g) are the major components of short channel effects in the devices [24]. The major leakage current components in the 6T SRAM cell during HOLD mode considering logic 0 and 1 are stored at the sensitive nodes Q and QB, respectively, as shown in Figure 6.5a, and are given by

$$I_{sub_{6T}} = I_{sub_{MLA}} + I_{sub_{MLP}} + I_{sub_{MRN}}, \tag{6.11}$$

$$I_{g_{6T}} = I_{gd_{MLA}} + I_{gd_{MLP}} + I_{gd_{MLN}} + I_{gs_{MLN}} + I_{gs_{MRP}} \\ + I_{gd_{MRP}} + I_{gd_{MRN}} + I_{gd_{MRA}} + I_{gs_{MRA}}, \tag{6.12}$$

$$I_{jn_{6T}} = I_{jnd_{MLA}} + I_{jnd_{MLP}} + I_{jnd_{MRN}} + I_{jnd_{MRA}} + I_{jns_{MRA}}, \tag{6.13}$$

$$I_{Leakage_{6T}} = I_{sub_{6T}} + I_{g_{6T}} + I_{jn_{6T}}. \tag{6.14}$$

Similarly, the major leakage current components in the proposed AS10T SRAM cell for the HOLD operation considering logic 0 and 1 are stored at the storage nodes Q and QB, respectively, as shown in Figure 6.5b, and are given by

$$I_{sub_{AS10T}} = I_{sub_{MLA}} + I_{sub_{MLP}} + I_{sub_{MRN}} + I_{sub_{MV2}}, \tag{6.15}$$

$$I_{g_{AS10T}} = I_{gd_{MLA}} + I_{gd_{MLP}} + I_{gd_{MLN}} + I_{gs_{MLN}} + I_{gs_{MRP}}$$
$$+ I_{gd_{MRP}} + I_{gd_{MRN}} + I_{gd_{MRA}} + I_{gs_{MRA}} + I_{gs_{MR2}} \quad (6.16)$$
$$+ I_{gd_{MR2}} + I_{gs_{MV1}} + I_{gd_{MV1}} + I_{gd_{MV2}},$$

$$I_{jn_{AS10T}} = I_{jnd_{MLA}} + I_{jnd_{MLP}} + I_{jnd_{MRN}} + I_{jnd_{MRA}}$$
$$+ I_{jns_{MRA}} + I_{jnd_{MV2}}, \quad (6.17)$$

$$I_{\text{Leakage}_{AS10T}} = I_{\text{sub}_{AS10T}} + I_{g_{AS10T}} + I_{jn_{AS10T}}. \quad (6.18)$$

From Figures 6.5a and 6.5b, we observe that the proposed AS10T SRAM cell has more leakage components as compared to the 6T SRAM cell. Most of the additional

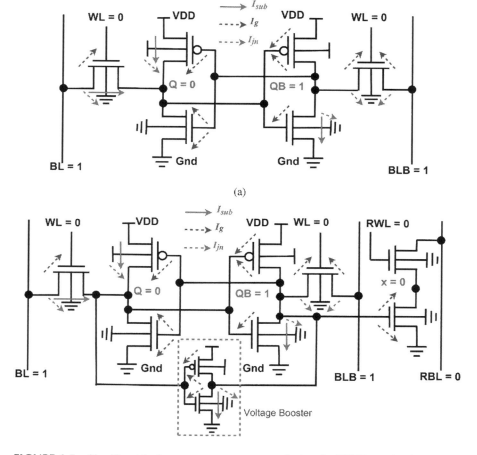

FIGURE 6.5 Significant leakage current components during the HOLD mode of operation considering logic 0 and 1 are stored at the storage nodes Q and QB for (a) a 6T SRAM cell and (b) an AS10T SRAM cell.

components are from I_g, which slightly increases the effective leakage power dissipation in the proposed circuit. All the additional components are because of the read-decoupled and voltage-booster circuit used in the proposed AS10T SRAM cell to improve the read stability and storage node critical charge.

6.5 SRAM BITCELL DESIGN METRICS

To study the performance of circuits considering a radiation-hardened design in combination with the aging of semiconductor devices, we refer to the model already available for the PTM 32-nm CMOS technology [1]. The simulations are performed using Synopsis HSPICE [26] considering a supply voltage $V_{DD} = 0.9$ V and the operating temperature is set to 25°C unless explicitly specified.

6.5.1 STABILITY ANALYSIS

For the stability analysis of the previously mentioned SRAM cells, we have calculated the HSNM, the RSNM, and the write margin (WM) at $V_{DD} = 0.9$ V, as shown in Figure 6.6a. The stability is conventionally computed as the static noise margin. The noise margin is gauged by tracing the overlapped VTC (butterfly diagram) for the back-to-back connected inverters that form the memory cell. The diagonal of the largest square that can fit in the eyes of the butterfly diagram finally determines the noise margin [21]. One can observe a similar WM for the 6T, RD8T, and AS10T SRAM cells, whereas the WM of AS8T is reduced by 2.1% compared to the others. The reduced WM is due to the charge booster that is connected between the storage nodes, which strengthens the node by increasing the node's capacitance. The HSNM of 6T and RD8T is the same because during the hold operation, the same transistor is active, whereas the HSNM of AS8T and AS10T is increased by 4.17%, referred to in the 6T SRAM cell. This increase in HSNM of AS8T and AS10T is due to the presence of the charge booster connected between the storage nodes. One drawback of the 6T SRAM cell is its small RSNM. The read-decoupled technique is one of the effective ways to increase RSNM because of the presence of a separate path during the read operation. The butterfly curve of 6T and AS10T is shown in Figure 6.6b, and clearly shows the RSNM of AS10T being higher than for the 6T SRAM cell. The RSNM of RD8T, AS8T, and AS10T is increased by 3.36×, 1.36×, and 3.29× compared to the 6T SRAM cell. The enhancement of the RSNM for RD8T and AS10T is due to the presence of a separate read-decoupled path during the read operation, whereas the RSNM for AS8T cell is high due to the charge-booster circuit.

6.5.2 POWER DISSIPATION ANALYSIS

The power dissipation of the SRAM cell is one of the major concerns for low-power circuit designs. We calculated the power dissipation of all the considered SRAM cells at the 0.9V supply voltage and 25°C for all three modes of the operations, namely, read, write, and hold modes, as shown in Figure 6.7. The results show that the read power for the circuits having a read-decoupled circuit increases, whereas the voltage booster decreases the power dissipation during the read operation. The

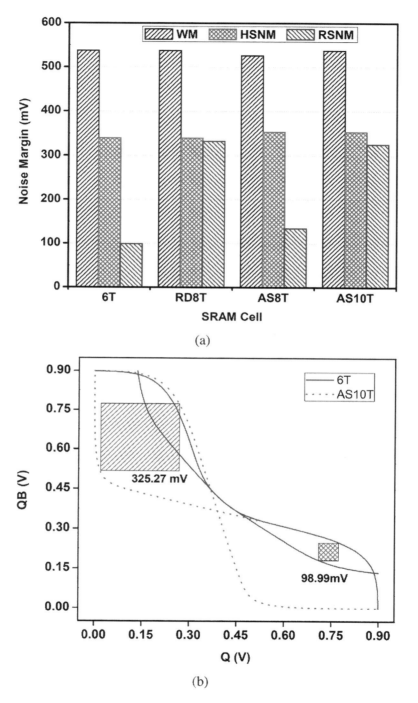

FIGURE 6.6 (a) WM, HSNM, and RSNM of different SRAM cells; (b) butterfly curve for read operation and RSNM of the 6T and AS10T SRAM cells. The margins are calculated considering a supply voltage of $VDD = 0.9$ V and $T = 25°C$.

BTI-Aware and Soft-Error-Tolerant SRAM 187

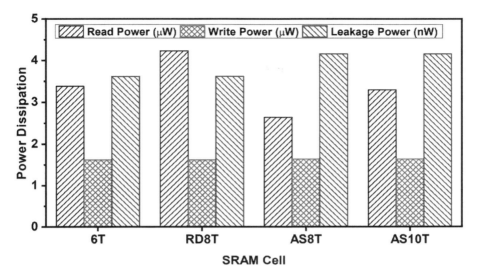

FIGURE 6.7 Power Dissipation of Various SRAM Cells during Read Operation, Write Operation, and Hold Operation.

proposed AS10T has both read decoupled and voltage booster circuits hence effectively less read power dissipation compared to the 6T SRAM cell. However, the read power dissipation for the RD8T and AS8T SRAM cells is less than the AS10T SRAM cell. Results also show that the power dissipation during the write operation for all the considered cells is almost the same. This is because, during the write operation, all the cells have the same cross-coupled inverters to write the data at the storage nodes. We also analyze the power during hold mode or leakage power of all the considered cells. Results demonstrate that the voltage booster increases the current paths, which leads to more leakage power dissipation in the AS8T and proposed AS10T SRAM cells. The components that are responsible for increasing the leakage power for AS10T have already been discussed in Section 6.4. The normalized values of all the power components can be seen in Table 6.7.

6.5.3 Delay Analysis

Read access time or read delay (T_{RA}) is the time duration between the read wordline (RWL) activation to the instant when the read bitline (RBL) voltage is discharged by 50mV from its initial logic high. For the differential read SRAM cell, the 50mV differential between bitline (*BL*) and bitline bar (*BLB*) is good enough to be detected by a sense amplifier, thereby avoiding a misread [16]. However, in the case of a single-ended read, T_{RA} is the time required for discharging the bitline voltage to (V_{DD}: 50mV) after the activation of WL during a read operation [19]. Similarly, the write access time or write delay (T_{WA}) for writing '1' is estimated as the time duration between the WL activation time to the time when a '0' storing node charges up to

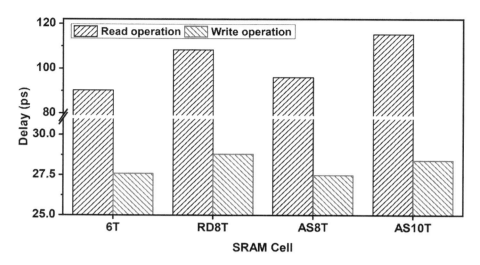

FIGURE 6.8 Delays of various SRAM cells during the read operation and the write operation.

90% of V_{DD}. Similarly, T_{WA} for writing '0' is estimated as the time duration between the WL activation time to the time when a '1' storing node discharges to 10% of V_{DD}.

Figure 6.8 shows the read and write delays for all the considered SRAM cells. The results show that the circuits with read-decoupled circuit has slightly more delay because the read-decoupled circuit increases the overall capacitance, which leads to increased delay. The effect of read-decoupled circuit on the write delay is not very high because it is not used during the write operation. However, the read-decoupled circuit affects the read delay due to its involvement during the read operation. The read delay of the proposed AS10T SRAM cell is 27.7% higher than the 6T SRAM cell.

6.5.4 Critical Charge Analysis

To estimate the critical charge at the storage nodes of the discussed SRAM cells, we assumed that either '0' or '1' is stored at storing nodes Q and QB, respectively. Since the carrier mobility of the NMOS transistor is higher than that for the PMOS transistor, the node storing '1' is more susceptible to any soft errors than the node storing '0' [17]. Therefore, we consider the node QB as the weakest node, being responsible for the effective critical charge Q_{crit} of the SRAM cells. Figure 6.9 shows the critical charge at the nodes Q and QB for the considered SRAM cells. It can be observed that the critical charge at node Q is higher than the critical charge at node QB, and the proposed AS10T cell exhibits the highest critical charge among all studied cells; that is, the critical charge of the AS10T is 75.83%, 75.41%, and 0.14% higher compared to the 6T, RD8T, and AS8T SRAM cells, respectively. The AS10T and AS8T have a higher critical charge because the charge booster is connected between the storage

BTI-Aware and Soft-Error-Tolerant SRAM 189

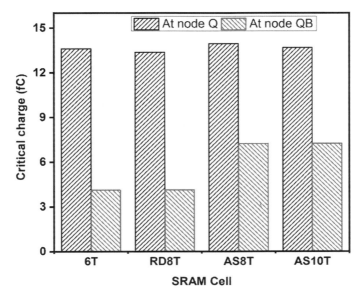

FIGURE 6.9 Critical charge at storage nodes of SRAM cells considering logic 0 and 1 are stored at nodes Q and QB, respectively.

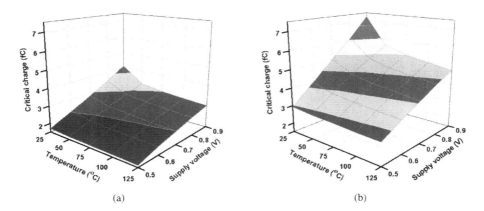

FIGURE 6.10 Critical charge variation with different supply voltages and temperatures for (a) a 6T SRAM cell and (b) an AS10T SRAM cell.

nodes. The charge booster increases the current path at the sensitive nodes, which increases the node capacitance and hence increases the soft-error tolerance. The AS10T behaves similarly to AS8T in terms of the critical charges of the most sensitive nodes. Based on the preceding observation, QB is considered the most sensitive node of the SRAM cells and is considered for further analysis.

Next, the critical charge is calculated for different supply voltages and device temperatures for 6T and AS10T SRAM cells, as shown in Figure 6.10. It can be observed that the critical charge for the AS10T cell is higher at each supply voltage and temperature. It has to be noted that the higher the critical charge, the better the hardening with respect to soft errors. Thus, the proposed AS10T cell exhibits enhanced soft-error hardening compared to the 6T SRAM cell and compared to the other cells. As can be further seen, the critical charge reduces with increasing temperature, whereas the critical charge increases at larger supply voltages. From the previous discussions, the critical charge change with temperature and supply voltage can be modeled as

$$Q_{\text{crit}} + \Delta Q_{\text{crit}} = \frac{k \times (V_{DD} + \Delta V_{DD})}{n \times (T + \Delta T)}, \qquad (6.19)$$

where k and n are technology and material-dependent constants. The parameter k depends linearly on the supply voltage, whereas n depends exponentially on the operating temperature. ΔQ_{crit} is the change in critical charge due to a change in supply voltage (ΔV_{DD}) and operating temperature (ΔT).

6.5.4.1 Effect of Rise Time and Fall Time

Furthermore, we also analyze the effect of SEU with different t_r and t_f on the sensitive node of the 6T and AS10T cells, as shown in Figure 6.11. Our results demonstrate that the critical charge mainly depends on the t_f because it determines the total charge stored at the sensitive node of the SRAM cells, as depicted in Figure 6.3.

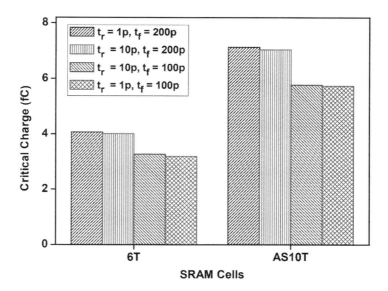

FIGURE 6.11 SEU effect on the observed critical charge of sensitive nodes of the SRAM cells simulated considering different rise times (t_r) and fall times (t_f) of the injected double exponential current pulse at the sensitive nodes.

BTI-Aware and Soft-Error-Tolerant SRAM

When the fall time t_f of the injected pulse increases, the charge collection probability also increases, which provides a higher critical charge at the sensitive nodes of the SRAM cells.

6.5.5 Aging Effects on SRAM Cells' Performance

As mentioned before, the performance of integrated MOS transistors can change over time [23]. In general, BTIs are considered responsible for the aging of the devices.

In the following, we analyze the stability of the SRAM cells and the evolution of the critical charge when a BTI is being considered a consequence of device stress.

6.5.5.1 On SRAM Stability

For the analysis of aging effects on the HSNM of the 6T and AS10T SRAM cells, we consider NBTI and PBTI during stress for PMOS and NMOS transistors, respectively. The DC stress is calculated to be applied for 3 years of stress time. The effect of BTIs on the 6T and AS10T SRAM stability for the hold operation is given in Table 6.2. The result shows that the HSNM reduces by 2.13% and 1.92% for the 6T and AS10T cells, respectively, considering only NBTI, and 4.26% and 3.85% for the 6T and AS10T cells, respectively, when considering both NBTI and PBTI. Results demonstrate that the AS10T is slightly less affected by BTIs compared to the 6T SRAM cell, which indicates that the AS10T cell is more stable in terms of BTIs.

6.5.5.2 On Critical Charge

To analyze the effect of BTIs on soft error susceptibility, we calculated the critical charge while considering the impact of BTIs for a stress time of 3 years. Table 6.3 shows the critical charge variation observed at the sensitive node of the 6T and AS10T SRAM cells with stress considering NBTIs and PBTIs. The result shows that the critical charge decreases with stress time. The critical charge of the 6T and AS10T is reduced by 6.39% and 0.84%, respectively, when considering only NBTIs. Similarly, the critical charge of the 6T and the AS10T is reduced by 7.86% and 0.98%, respectively, when again considering only NBTIs. This indicates that the critical charge of AS10T is slightly less affected by the BTIs as compared to the 6T SRAM cell, and thus, the proposed cell is more robust against aging effects.

TABLE 6.2
HSNM of Different SRAM Cells with and without BTIs after the Stress Time of 3 Years

Stress	6T	AS10T
No stress	332.3 mV	367.6 mV
NBTI	325.2 mV	360.6 mV
NBTI + PBTI	318.2 mV	353.5 mV

TABLE 6.3
Critical Charge (fC) at the Sensitive Node of the Respective SRAM Cells with and without BTIs after 3 Years of Stress Time for BTIs for both NBTI/PMOS and PBTI/NMOS Occurring in Transistors

Stress time (years)	6T NBTI	6T BTI	AS10T NBTI	AS10T BTI
0	4.07	4.07	7.12	7.12
1	3.88	3.83	7.08	7.07
2	3.84	3.78	7.07	7.06
3	3.81	3.75	7.06	7.05

6.5.6 SOFT ERROR ANALYSIS

To analyze the soft error robustness of SRAM cells the analysis of SERR is needed for all the SRAM cells.

6.5.6.1 Inter-Cell SERR

The intra-cell SERR with both supply voltage and temperature change, namely $SERR_V$ and $SERR_T$, respectively, are analyzed in this section. Table 6.4 shows the SERR with supply voltage change ($SERR_V$) from 0.5V to 0.9V at different operating temperatures. From the results, it is observed that the $SERR_V$ increases with the rise in temperature for all the considered SRAM cells. However, the increase in $SERR_V$ for the proposed AS10T SRAM cell is less compared to all other considered cells. The change in $SERR_V$ with a change in temperature from 25°C to 125°C for the proposed AS10T SRAM cell is 0.112 as compared to 0.223, 0.221, and 0.113 for the 6T, RD8T, and AS8T SRAM cells, respectively. Hence, the proposed cell has less variation in $SERR_V$ as compared to all other cells. Similarly, Table 6.5 shows the SERR with an operating temperature change ($SERR_T$) from

TABLE 6.4
SERR with Voltage Change ($SERR_V$) at Various Operating Temperatures of Different SRAM Cells

Temperature	SRAM cells ↓			
	6T	RD8T	AS8T	AS10T
25°C	0.089	0.089	0.016	0.014
50°C	0.217	0.215	0.063	0.061
75°C	0.257	0.254	0.089	0.087
100°C	0.284	0.282	0.107	0.105
125°C	0.313	0.311	0.129	0.126

TABLE 6.5
SERR with Temperature Change ($SERR_T$) at Various Supply Voltages of Different SRAM Cells

Supply voltage	SRAM cells ↓			
	6T	RD8T	AS8T	AS10T
0.5V	0.903	0.899	0.785	0.782
0.6V	0.719	0.701	0.498	0.496
0.7V	0.553	0.551	0.298	0.295
0.8V	0.406	0.402	0.170	0.167
0.9V	0.262	0.260	0.086	0.082

25°C to 125°C at different supply voltages. From the results, it is observed that the $SERR_T$ decreases with the rise in supply voltage for all the considered SRAM cells. However, the $SERR_T$ for the proposed AS10T SRAM cell is less compared to all other considered cells. The decrease in $SERR_T$ with change in the supply voltage from 0.5V to 0.9V for the proposed AS10T SRAM cell is 0.7 as compared to 0.641, 0.639, and 0.699 for the 6T, RD8T, and AS8T SRAM cells, respectively. Hence, the proposed cell is more resilient to soft error at a higher supply voltage compared to all other cells.

6.5.6.2 Inter-Cell SERR

The analysis of inter-cell SERR for the effectiveness of the proposed SRAM cell on the soft-error hardening as compared to the 6T SRAM cell is also needed. Figure 6.12 shows the inter-cell SERR for all the considered cells normalized to the 6T SRAM cell for different extreme combinations of supply voltage and operating temperature. As we have two variables, V_{DD} and T for calculating inter-cell SERR, there is a total of four combinations. Here, we have calculated SERR considering supply voltages of 0.5V and 0.9V, and operating temperatures 25°C and 125°C. The results show that the inter-cell SERR for RD8T is nearly the same as the 6T SRAM cell because of the equal critical charges at the sensitive nodes. The inter-cell SERR for the AS8T and the proposed AS10T are also the same because the voltage booster is connected between storage nodes to improve the critical charge in both the cells but less than the 6T and RD8T SRAM cells. It is also observed that the inter-cell SERR also depends on the supply voltage and the operating temperature. The results show that the maximum improvement in the SERR is when the operating temperature is 25°C, and the supply voltage is 0.9V. Higher supply voltage increases the current flowing through the circuit and leads to an increase in the critical charge at the sensitive node. Similarly, the critical charge decreases with the increase in operating temperature; hence, the SERR at the higher temperature increases. From the preceding discussions, the proposed SRAM cell performs well in terms of soft-error hardening if it is operated at room temperature and subthreshold conditions.

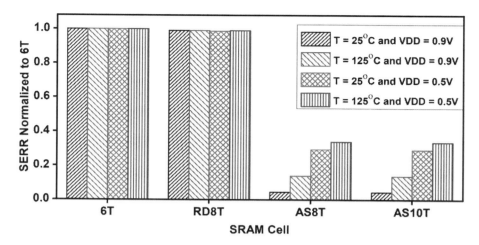

FIGURE 6.12 Inter-cell SERR for different SRAM cells at the four extreme combinations of supply voltage and operating temperature. The SERR is normalized to the 6T SRAM cell for all the combinations.

6.5.7 Effect of Supply Voltage and Temperature Variation Analysis

The impact of supply voltage and temperature variations on the soft-error tolerance and energy-efficient circuit is an important measure for the stable operation and low-power application of the circuit.

6.5.7.1 On Critical Charge

The effect of supply voltage and temperature variations on the critical charge of sensitive node QB is shown in Figure 6.13. Figure 6.13a shows the evolution of the critical charge with supply voltage variation in the range of $V_{DD} = 0.8V$ to $1V$ for the considered circuits. The results reveal that the critical charge increases with the supply voltage as the node capacitance increases with the same. It is also observed that the critical charge for the proposed AS10T is higher compared to the 6T SRAM cell, indicating that the proposed AS10T cell is more hardened to soft errors caused by supply variations. Figure 6.13b shows the trend of the critical charge with the temperature ranging from $T = 25°C$ to $T = 125°C$. The results show that with an increase in temperature, the critical charge decreases because the device carrier mobility decreases.

6.5.7.2 On Power Consumption

The power consumption during hold operation with supply and temperature variation for 6T and AS10T SRAM cells is shown in Figure 6.14. For our analysis, we considered the supply voltage ranging from $V_{DD} = 0.8V$ to $1V$ and the operating temperature ranging from $T = 25°C$ to $T = 125°C$. The power dissipation increases with increasing supply voltage and increasing temperature. The results show that the

BTI-Aware and Soft-Error-Tolerant SRAM

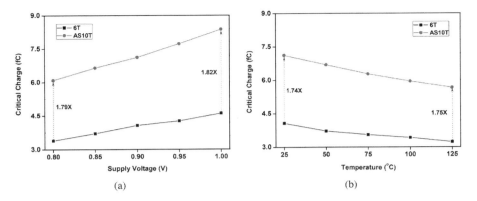

FIGURE 6.13 Critical charge at the sensitive node of SRAM cells (a) with supply voltage variation (b) with temperature variation.

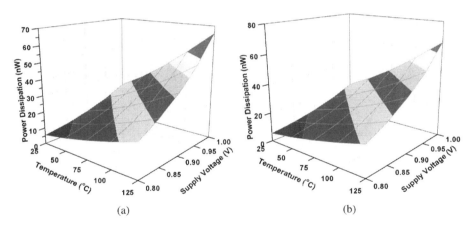

FIGURE 6.14 Power consumption with supply voltage and temperature variation for the (a) 6T SRAM cell and the (b) AS10T SRAM cell.

AS10T cell has less power variation with supply voltage and temperature as compared to the 6T SRAM cell.

6.5.8 Process Variation Analysis

For the investigation of process variations on the critical voltage of considered SRAM cells implemented using deep-submicron technologies, both inter-die and intra-die variation needs to be analyzed. The inter-die variation is caused by the deviation in the photolithographic process and parameter variation on the same device manufactured at varying time over a larger area. It is systematic in nature and affects all devices equally in a die. In contrast to the systematic inter-die variation, this variation is due to parameter mismatches across the two identical devices placed next to each other on the same die [18]. The intra-die variation is unpredictable in nature and is caused by random uncertainties in the fabrication process, such as variations in oxide thickness, length, width, number of dopants, and flat band control, among others [8].

6.5.8.1 Inter-Die Variations

The critical charge for the 6T and AS10T cells for the process corners (fast–fast [FF], fast–slow [FS], slow–fast [SF], slow–slow [SS], typical–typical [TT]), as shown in Figure 6.15. The results show that the critical charge for the AS10T is higher than that of the 6T cell for all the process corners. The previous observation indicates that the AS10T SRAM cell is more robust to the soft-error tolerance.

6.5.8.2 Intra-Die Variations

Additionally, for the analysis of soft-error resilience and process variations on the 6T and AS10T SRAM cells, we perform 5000 Monte Carlo simulations for the critical voltages of the storage node considering the variations in the threshold voltage of the transistors. The threshold voltage of the transistors is generated randomly using a normal distribution with ±10% maximum deviations from its original value [17].

Figure 6.16 shows the simulation results for the 5,000 sample of critical voltage of the 6T and AS10T SRAM cells. The simulation result shows that the AS10T SRAM cell has less effect of process variations as compared to the 6T SRAM cell. The

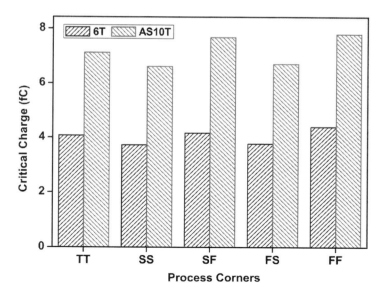

FIGURE 6.15 Critical charge at the sensitive node of SRAM cells for different process corners.

BTI-Aware and Soft-Error-Tolerant SRAM

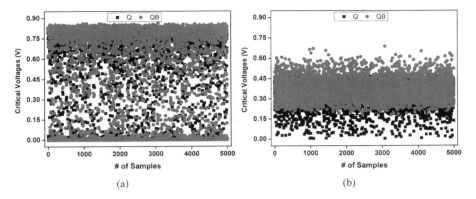

FIGURE 6.16 Distribution of critical voltages (V_{crit}) with 5000 Monte Carlo samples for V_Q and V_{QB} for (a) 6T SRAM cell and (b) AS10T SRAM cell. The distribution indicates the repeatability of the critical voltage for 6T and AS10T SRAM cells.

results demonstrate that a significant portion of critical voltage repeatability for the 6T cell is near 0V, which indicates that the flipping of logic in these SRAM cells can be triggered earlier as compared to the AS10T SRAM cell.

Furthermore, we calculated the standard deviation considering ±3σ deviation from the mean (µ) value. The calculation formula of standard deviation (σ) for the critical voltage of the SRAM cells is given as

$$\sigma = \sqrt{\frac{\sum (X_i - \mu)^2}{N}}, \qquad (6.20)$$

where σ is the standard deviation (SD) and N, X_i, and μ denote the number of sample values, that is, 5,000, the sample value, and the mean value, respectively.

For the analysis of soft-error resilience and process variations, a Gaussian distribution for the storage node critical voltages of 6T, RD8T, AS8T, and AS10T SRAM cells is shown in Figure 6.17. The figure shows the distribution of critical voltages for each sample of various SRAM cells. The simulation results show that the proposed AS10T SRAM cell has less effect on process variations compared to all other cells. The results demonstrate that the significant portion of critical voltage repeatability for 6T and AS8T is near 0V, which shows that the flipping of logic in these SRAM cells is easier compared to RD8T and AS10T SRAM cells. The scattered response of critical voltage for the SRAM cells represents the sensitivity of process variations that reduces the reliability of the circuit. Table 6.6 shows the mean (µ) and standard deviation (σ) of the storage node critical voltages by performing 5,000 Monte Carlo simulations for all the considered SRAM cells. The proposed cell has deviations of 0.088V and 0.077V for Q and QB, respectively, whereas the 6T SRAM cell has deviations of 0.356V and 0.364V for Q and QB, respectively. The results demonstrate that the deviation of critical voltage from the mean value for the proposed AS10T

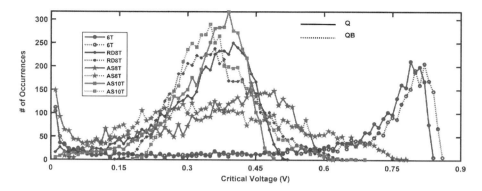

FIGURE 6.17 5,000 Monte Carlo–simulated Gaussian plot of V_Q and V_{QB} critical voltages (V_{crit}) for different SRAM cells.

TABLE 6.6
Mean and Standard Deviation of Storage Node Critical Voltages (in volts) for Different SRAM Cells

SRAM cells ↓	V_{crit} @Q		V_{crit} @QB	
	μ	σ	μ	σ
6T	0.376	0.356	0.379	0.364
RD8T	0.334	0.103	0.364	0.088
AS8T	0.323	0.171	0.378	0.163
AS10T	0.336	0.088	0.366	0.077

SRAM cell is minimum as compared to all other considered SRAM cells. From results, it is also observed that the 6T SRAM cell is maximum affected by process variations, whereas the proposed AS10T SRAM cell has a better tolerance of process variations.

Furthermore, we also evaluated the process **variability** for the critical voltages of all the considered SRAM cells. The process variability is the ratio of the standard deviation and the mean value of the critical voltages. For better tolerance of the process variations, the variability should be as small as possible [25]. Figure 6.18 shows the variability for the critical voltages of storage node Q and QB for all the considered SRAM cells, and we observe that the AS10T has less variability as compared to all other SRAM cells. Furthermore, our results demonstrate that the variability at node V_Q for AS10T is 3.61×, 1.18×, and 2.02× lower than the 6T, RD8T, and AS8T SRAM cells, respectively. Similarly, the variability at node V_{QB} for AS10T is 4.56×, 1.15×, and 2.05× lower than for 6T, RD8T, and AS8T SRAM cells, respectively.

BTI-Aware and Soft-Error-Tolerant SRAM

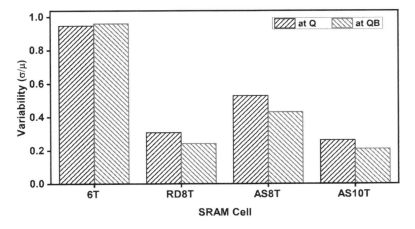

FIGURE 6.18 Process variability (σ/μ) of V_Q and V_{QB} critical voltages (V_{crit}) for different SRAM cells.

6.5.9 Area Comparison

Cell area is one major concern when designing energy-efficient, reliable and high-performance SRAM cell. Figure 6.19 shows the layout view of the 6T and proposed AS10T SRAM cells. Both of the cells have two sensitive nodes as indicated in the figure. Out of the two sensitive nodes, one of the nodes is more sensitive to soft error

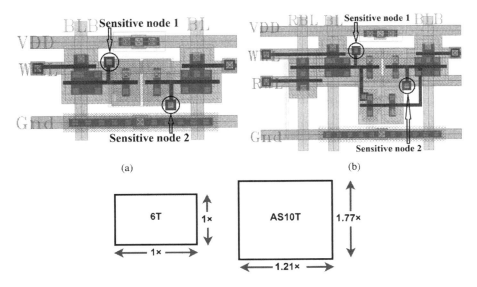

FIGURE 6.19 Layout of (a) the 6T SRAM cell and (b) the proposed AS10T SRAM cell.

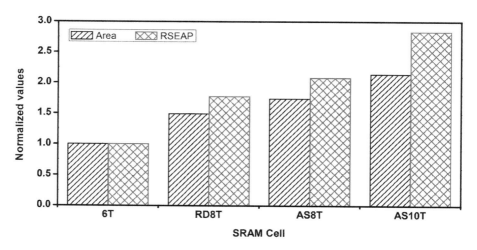

FIGURE 6.20 Normalized values of cell area and reliable stability to energy area product for different SRAM cells.

depending on the stored data. Figure 6.20 shows the normalized area of different SRAM cells. The area of RD8T, AS8T, and AS10T are 1.49×, 1.74×, and 2.14× higher, respectively, compared to the 6T SRAM cell.

6.5.10 Reliability and Stability–to–Energy Area Product Ratio

The performance parameters of SRAM cells have a trade-off among them [23]. Therefore, to access the performance of SRAM cells, we propose a reliability stability to energy area product (**RSEAP**) ratio to evaluate the overall novelty of the SRAM cell as given by:

$$RSEAP = \frac{Q_{\text{crit,n}} \times RSNM_n \times HSNM_n \times WM_n}{P_{\text{W,n}} \times T_{\text{WA,n}} \times P_{\text{R,n}} \times T_{\text{RA,n}} \times P_{\text{L,n}} \times A_n}, \qquad (6.21)$$

where Q_{crit},n, $RSNM_n$, $HSNM_n$, WM_n, $P_{\text{W,n}}$, $P_{\text{R,n}}$, $P_{\text{L,n}}$, $T_{\text{WA,n}}$, $T_{\text{RA,n}}$, and A_n are the normalized values of critical charge, read SNM, hold SNM, write margin, write power, read power, leakage power, write delay, read delay, and cell area, respectively. All the parameters are normalized with respect to 6T SRAM cell. Figure 6.20 shows the normalized RSEAP at 0.9V supply voltage of different SRAM cells considered in this work. From the results, it is observed that the normalized RSEAP of the proposed AS10T SRAM cell is the highest among all the considered cells. The RSEAP of RD8T, AS8T, and AS10T are 1.77×, 2.08×, and 2.83× higher than the 6T SRAM cell, respectively. The normalized values of all the performance parameters of considered cells are given in Table 6.7. It is also observed that the proposed cell has a higher critical charge and noise margin. Therefore, the AS10T cell is an alternative option when considering overall performance with the enhanced soft-error hardening.

TABLE 6.7
Actual and Normalized Value of Various Performance Parameters for Different SRAM Cells

Performance parameters ↓	SRAM Cells ↓			
	6T	RD8T	AS8T	AS10T
Critical charge (fC)	4.121	4.131	7.236	7.246
Normalized Critical charge	1	1.002	1.756	1.758
WM (mV)	537.40	537.40	526.08	537.40
Normalized WM	1	1.000	0.979	1.000
HSNM (mV)	339.41	339.41	353.55	353.55
Normalized HSNM	1	1.000	1.042	1.042
RSNM (mV)	98.99	332.34	134.35	325.27
Normalized RSNM	1	3.357	1.357	3.286
Read power (µW)	3.38	4.23	2.64	3.30
Normalized read power	1	2.201	1.187	2.201
Write power (µW)	1.62	1.62	1.64	1.64
Normalized Write power	1	1.001	1.014	1.014
Leakage power (nW)	13.61	3.62	4.16	4.16
Normalized leakage power	1	1.001	1.151	1.151
Read delay (ps)	90.29	108.35	96.11	115.33
Normalized read delay	1	1.943	1.721	1.926
Write delay (ps)	27.59	28.80	27.48	28.40
Normalized write delay	1	1.044	0.996	1.029
Normalized area	1	1.49	1.74	2.14
RSEAP	1	2.16	1.21	2.34

6.6 SUMMARY

This chapter presents a radiation-hardened asymmetric 10T (AS10T) SRAM cell to enhance the soft-error hardening, considering aging effects like bias temperature instabilities. The proposed cell uses a read-decoupled path to improve the RSNM and a voltage booster connected between the storage nodes to improve node capacitance and finally lead to enhanced radiation hardening. We demonstrated that the AS10T SRAM cell has a higher critical charge at the most sensitive node compared to other SRAM cells. Furthermore, we analyzed the effect of BTIs on soft-error susceptibility on the critical charge and on the stability of the SRAM cells and observed that the AS10T cell is more resilient to aging effects. Therefore, we conclude that the proposed SRAM cell can be used for applications, which demand not only high speed but especially also a radiation-hardened design and high operating stability with considerable chip area.

6.7 GLOSSARY

Reliability stability to energy area product (RSEAP): This is the ratio of the product of normalized critical charge and different noise margins to the normalized value of power delay and area product.

Single-event transient: It is a transient fault that can act as a electrical pulse and propagate through combinational logic paths.
Single-event upset: It is a change of state caused by one ionizing particle (ion, neutron, photon) striking a sensitive node in a storage element.
Soft error: A soft error occurs when a radiation event causes enough of a charge distribution to reverse or flip the data state of a memory cell, latch, flip-flop, or even a node in a combinational block.
Soft-error rate ratio: The ratio of soft error rate for the considered circuit to the soft error rate of reference circuit.
Transient fault: The high-energy particle affects the performance of the transistor and generates a glitch at the output of the circuit. This erroneous behavior is referred to as a transient fault.
Variability: It is the ratio of standard deviation (σ) and mean value (μ).

6.8 QUESTIONS

1. What is the critical charge?
2. What is the critical time and critical voltage while analyzing soft error in the SRAM cell?
3. How to identify the most sensitive node in the SRAM cell?
4. Which major leakage current is generally considered for the power dissipation analysis in SRAM cells?
5. What are the various parameters that decide the critical charge at the sensitive node of an SRAM cell?
6. What is the difference between NBTI and PBTI? Which one is more dominant in the circuit?
7. What is the difference between single-event transient and single-event upset?
8. What is the difference between inter-die and intra-die process variations?
9. What is meant by variability?

REFERENCES

1. Predictive Technology Model (PTM). https://ptm.asu.edu/. Accessed on December 12, 2020.
2. Sayeed Ahmad, Naushad Alam, and Mohd Hasan. Pseudo differential multi-cell upset immune robust SRAM cell for ultra-low power applications. *AEU-International Journal of Electronics and Communications*, 83:366–375, 2018.
3. Ihsen Alouani, Wael M Elsharkasy, Ahmed M Eltawil, Fadi J Kurdahi, and Smail Niar. AS8-static random access memory (SRAM): Asymmetric SRAM architecture for soft error hardening enhancement. *IET Circuits, Devices & Systems*, 11(1):89–94, 2017.
4. Leland Chang, David M Fried, Jack Hergenrother, Jeffrey W Sleight, Robert H Dennard, Robert K Montoye, Lidija Sekaric, Sharee J McNab, Anna W Topol, Charlotte D Adams, et al. Stable SRAM cell design for the 32 nm node and beyond. In *VLSI Technology, 2005. Digest of Technical Papers. 2005 Symposium on*, pages 128–129. IEEE, 2005.
5. Qian Ding, Rong Luo, Hui Wang, Huazhong Yang, and Yuan Xie. Modeling the impact of process variation on critical charge distribution. In *SOC Conference, 2006 IEEE International*, pages 243–246. IEEE, 2006.

6. Paul E Dodd and Lloyd W Massengill. Basic mechanisms and modeling of single-event upset in digital microelectronics. *IEEE Transactions on Nuclear Science*, 50(3):583–602, 2003.
7. Veronique Ferlet-Cavrois, Lloyd W Massengill, and Pascale Gouker. Single event transients in digital CMOS: A review. *IEEE Transactions on Nuclear Science*, 60(3):1767–1790, 2013.
8. Swaroop Ghosh and Kaushik Roy. Parameter variation tolerance and error resiliency: New design paradigm for the nanoscale era. *Proceedings of the IEEE*, 98(10):1718–1751, 2010.
9. Thomas Granlund, Bo Granbom, and Nils Olsson. Soft error rate increase for new generations of SRAMs. *IEEE Transactions on Nuclear Science*, 50(6):2065–2068, 2003.
10. Jing Guo, Lei Zhu, Yu Sun, Huiliang Cao, Hai Huang, Tianqi Wang, Chunhua Qi, Rongsheng Zhang, Xuebing Cao, Liyi Xiao, et al. Design of area-efficient and highly reliable RHBD 10T memory cell for aerospace applications. *IEEE Transactions on Very Large Scale Integration (VLSI) Systems*, 26(5):991–994, 2018.
11. David F Heidel, Kenneth P Rodbell, Phil Oldiges, Michael S Gordon, Henry HK Tang, Ethan H Cannon, and Cristina Plettner. Single-event-upset critical charge measurements and modeling of 65 nm silicon-on-insulator latches and memory cells. *IEEE Transactions on Nuclear Science*, 53(6):3512–3517, 2006.
12. HL Hughes and JM Benedetto. Radiation effects and hardening of MOS technology: Devices and circuits. *IEEE Transactions on Nuclear Science*, 50(3):500–521, 2003.
13. Eishi Ibe, Hitoshi Taniguchi, Yasuo Yahagi, Ken-ichi Shimbo, and Tadanobu Toba. Impact of scaling on neutron-induced soft error in SRAMs from a 250 nm to a 22 nm design rule. *IEEE Transactions on Electron Devices*, 57(7):1527–1538, 2010.
14. Shah M Jahinuzzaman, Mohammad Sharifkhani, and Manoj Sachdev. An analytical model for soft error critical charge of nanometric SRAMs. *IEEE Transactions on Very Large Scale Integration (VLSI) Systems*, 17(9):1187–1195, 2009.
15. Jin Sang Kim, Ik Joon Chang, et al. We-quatro: Radiation-hardened SRAM cell with parametric process variation tolerance. *IEEE Transactions on Nuclear Science*, 64(9):2489–2496, 2017.
16. Jaydeep P Kulkarni, Keejong Kim, and Kaushik Roy. A 160 mV robust Schmitt trigger based subthreshold SRAM. *IEEE Journal of Solid-State Circuits*, 42(10):2303–2313, 2007.
17. Sheng Lin, Yong-Bin Kim, and Fabrizio Lombardi. Design and analysis of a 32 nm PVT tolerant CMOS SRAM cell for low leakage and high stability. *Integration, the VLSI Journal*, 43(2):176–187, 2010.
18. Y Ohnari, AA Khan, A Dutta, M Miura-Mattausch, and HJ Mattausch. Die-to-die and within-die variation extraction for circuit simulation with surface-potential compact model. In *Microelectronic Test Structures (ICMTS), 2013 IEEE International Conference on*, pages 146–150. IEEE, 2013.
19. Ghasem Pasandi and Sied Mehdi Fakhraie. A 256-kb 9T near-threshold SRAM with 1k cells per bitline and enhanced write and read operations. *IEEE Transactions on Very Large Scale Integration (VLSI) Systems*, 23(11):2438–2446, 2014.
20. Daniele Rossi, Martin Omaña, Cecilia Metra, and Alessandro Paccagnella. Impact of bias temperature instability on soft error susceptibility. *IEEE Transactions on Very Large Scale Integration (VLSI) Systems*, 23(4):743–751, 2015.
21. Prachi Sanvale, Neha Gupta, Vaibhav Neema, Ambika Prasad Shah, and Santosh Kumar Vishvakarma. An improved read-assist energy efficient single ended PPN based 10T SRAM cell for wireless sensor network. *Microelectronics Journal*, 92:104611, 2019.
22. Ambika Prasad Shah, Nandakishor Yadav, Ankur Beohar, and Santosh Kumar Vishvakarma. An efficient NBTI sensor and compensation circuit for stable and reliable SRAM cells. *Microelectronics Reliability*, 87:15–23, 2018.

23. Ambika Prasad Shah, Nandakishor Yadav, Ankur Beohar, and Santosh Kumar Vishvakarma. On-chip adaptive body bias for reducing the impact of NBTI on 6T SRAM cells. *IEEE Transactions on Semiconductor Manufacturing*, 31(2):242–249, 2018.
24. Ambika Prasad Shah, Nandakishor Yadav, Ankur Beohar, and Santosh Kumar Vishvakarma. Process variation and NBTI resilient Schmitt trigger for stable and reliable circuits. *IEEE Transactions on Device and Materials Reliability*, 1–9, 2018.
25. Vishal Sharma, Maisagalla Gopal, Pooran Singh, and Santosh Kumar Vishvakarma. A 220 mV robust read-decoupled partial feedback cutting based low-leakage 9T SRAM for internet of things (IoT) applications. *AEU-International Journal of Electronics and Communications*, 87:144–157, 2018.
26. Synopsys. *Hspice User Guide: Simulation and Analysis*, 2010. https://www.synopsys.com/.
27. Aibin Yan, Zhengfeng Huang, Maoxiang Yi, Xiumin Xu, Yiming Ouyang, and Huaguo Liang. Double-node-upset-resilient latch design for nanoscale CMOS technology. *IEEE Transactions on Very Large Scale Integration (VLSI) Systems*, 25(6):1978–1982, 2017.

Index

A

access transistor, 3, 6, 47
address decoder, 81
aging, 18
analog, 10, 41, 68
ASIC, 1, 31

B

basic logic element (BLE), 31
beta ratio, 51, 54
bias temperature instability (BTI), 20, 176
bitline, 3, 13, 47
bitline capacitance, 14
bitline write trip point, 65
body bias sense amplifier (BBSA), 141
body effect coefficient, 8

C

cache memory, 4, 31, 35, 45
cell current, 84–85
cell ratio, 66
column decoder, 6, 47, 67, 73, 81
critical charge, 25, 177, 188, 190, 194
cross coupled inverter, 6, 46, 55, 125, 127
current feed sense amplifier (CFSA), 159, 165, 169, 170
current latch sense amplifier (CLSA), 125, 128, 129
current mode approach, 130
current sense amplifier, 154

D

data retention, 3, 55, 62, 98
data retention voltage, 12, 55
decoder, 6, 47, 67, 81
delay, 3, 19, 31, 37, 73, 77, 84, 87, 88, 92, 93, 99, 104, 109, 114, 120, 126, 127, 129–131, 142, 145–152, 165, 167, 171
digital, 1–3, 6, 10, 24, 31, 36, 68, 76, 90, 94, 115, 126, 141
DRAM, 45, 76
driver, 10, 48, 60, 68, 70
dynamic NBTI, 21

dynamic power dissipation, 3, 7, 73, 88, 111
dynamic read margin, 79, 87, 102, 113
dynamic read stability, 65
dynamic write stability, 67

E

electromigration, 18
embedded, 1 , 3, 4, 6, 35, 46
environmental variation, 19

F

FD-SoI, 28
Fermi potential, 20
field programmable gate array (FPGA), 31, 35, 77, 90
FinFET, 27, 154

G

glitching power, 7
global sense amplifier, 156

H

heterojunction TFET, 30
hold static noise margin (HSNM), 9, 13, 86
hot carrier injection (HCI), 18
hybrid sense amplifier, 154

I

input referred offset, 131, 142, 168
inter-cell SERR, 192
inter-die variation, 10, 16, 196
intra-die variation, 11, 16, 17, 196
IR drop, 19

L

latching in SRAM, 55
latency, 73
leakage current estimation of AS10T, 183–185
leakage power in SRAM, 71
look up table (LUT) of FPGA, 31
Low Power SRAM, 4–6

205

M

Monte Carlo, 84, 91, 93, 127, 128, 141, 143, 146, 162, 196–198
Moore's law, 1, 76, 132

N

N-curve, 59
negative BTI (NBTI), 176
noise margins, 55–56

O

object detection and tracking, 115–122
offset analysis, 168
offset analysis of SCSA, 140
offset condition CLSA, 136

P

performance analysis of sense amplifiers, 145
performance and power measurement of SA's, 171–173
PFC10T SRAM, 78, 80
PFC8T application in object tracking, 114
PFC8T SRAM, 105
positive BTI (PBTI), 176
power consumption in SRAM, 6
process, 6, 8, 10, 12, 14–17, 26, 28, 29, 33, 37
process variation analysis on SRAM, 195
PT10T SRAM, 94

R

radiation hardened 10T (AS10T) SRAM, 182
radiation hardening analysis, 177

RD Model, 20, 23
read driver, 68
read failure, 11
read margin, 51
read operation, 48–50
read yield issue, 132
reliability failure mechanisms, 15
reliability stability to energy area product (RSEAP), 200
Robust Subthreshold 8T SRAM for Image Processing, 103
RSNM, 60

S

scaling, 1, 3, 7, 11, 13, 15, 18, 24, 26, 28, 29, 33, 36, 71, 72, 76, 94, 132
sense amplifier, 10, 11, 13, 14, 27, 46, 47, 73, 79, 96, 97, 125–170, 187
silicon, 1, 20, 26–28
silicon on insulator, 27
SoC, 1, 3, 28, 33, 35, 76, 103
soft error rate, 25, 179, 180, 202
subthreshold, 3, 6–9, 13, 19, 28–30, 32–33, 35–37, 55
switching, 7, 15, 18, 19, 27, 50, 53, 68, 72, 127, 141, 147, 163, 164, 176

T

temperature, 15
temporal, 15, 18
time-dependent dielectric breakdown, 18

U

ultra-low-power, 3, 4, 77, 94